ON THE
PULSAR

ON THE
PULSAR

B. B. Kadomsev
formerly of Russian Academy of Sciences

Foreword by

A. Nomerotsky

World Scientific

NEW JERSEY · LONDON · SINGAPORE · BEIJING · SHANGHAI · HONG KONG · TAIPEI · CHENNAI

Published by

World Scientific Publishing Co. Pte. Ltd.

5 Toh Tuck Link, Singapore 596224

USA office: 27 Warren Street, Suite 401-402, Hackensack, NJ 07601

UK office: 57 Shelton Street, Covent Garden, London WC2H 9HE

Managing Editor of Translation: M. S. Aksenteva

Library of Congress Cataloging-in-Publication Data
Kadomtsev, B. B.
  [Na pul'sare. English]
  On the pulsar / B.B. Kadomsev.
    p. cm.
  ISBN-13 978-981-4289-72-6
  ISBN-10 981-4289-72-8
  1. Pulsars. 2. Neutron stars. 3. Astrophysics. I. Title.
  QB843.P8K3413 2009
  523.8'874--dc22
                                        2009026489

British Library Cataloguing-in-Publication Data
A catalogue record for this book is available from the British Library.

Typeset by Stallion Press
Email: enquiries@stallionpress.com

Printed in Singapore.

Contents

Foreword

This book about pulsars, by Academician Kadomtsev, may be surprising to people in the future because of the insights it provides into the field. He has combined an excellent introduction to pulsar physics with reflections on important informational and biological issues, all the while fusing this into a personal approach to educating the reader.

At first, the book appears to be a 'pop' book on physics but only for a few moments. The book contains two main strands: the first concerning the physics and the second dealing with a narrative about two boys that an old professor takes for a journey into the unknown. The story evolves in using the dialogues of Professor Leonid Andreevich with two teenagers who have a budding interest in science, Sasha and Misha. At times the story sounds like a piece of fiction, it has recollections from childhood, philosophical deviations and personal comments. Some of these tales are an interesting reflection on the period in which Kadomstev was writing the book at the end of last century, when Russia was going through an historic transition from socialism to capitalism.

Pulsars are the product of neutron stars, so this topic explains in close detail the behaviour of matter at extremely high density, when all nucleons are close together, thus creating a strong magnetic field. Professor Kadomtsev likes to start his explanations with simple things understandable by everybody, such as bouncing balls.

He then continues explaining the strange world of super strong magnetic fields, where mass is anisotropic and hydrogen atoms could form polymer chains.

Parts of the book are science fiction but they were deliberately designed so as to better illustrate the point in reference. In the beginning the reader will be able to tell the difference between that which is real and that which is fantasy but in the last two chapters the lines will not be so evident. There, the book takes an unusual turn and takes the reader to the wonders of self-organization and living organisms, topics which seem to be completely unrelated to the pulsar. The meaning of it becomes clear in the last section. Could the pulsar be sentient? To find out, the professor takes the boys to the pulsar itself and they find some evidence of self-organization, only to wake up to find it was a dream!

The level of the book is ambiguous. The author is driven by common sense and tries his best to make it an easy read. Kadomtsev believes that much of the book should be comprehensible to high school students. But, as a book written in Russia where the physics programme in schools and universities was, at the time, more advanced, the book therefore deals with topics which will not be familiar to a high school student in the US or Europe. The author tries to stretch the reader, to involve them in the process. His examples are highly original and revealing. The switches between simple and complicated topics occur very naturally during the dialogues; "what is the colour of an electron?" one of the students ask, "it is white", the professor replies. This would never occur to me!

Appendices explain the more complex subjects and have an excellent collection of problems with solutions.

This book has an unforgettable spirit which may seem old-fashioned to some or too 'soviet' to others. We can see Kadomtsev building a temple of science in front of us, while he explores different halls and chambers, some of them well illuminated and other slightly murkier. The voices of people like Landau and Sakharov still resound in this temple. The book clearly reflects the scientific and educational establishment of the Soviet Union during its last decades. This is what it was, and it is a system that had shortcomings, but at the same time produced scores of the brightest minds.

So go on, reader, explore and enjoy this book on the pulsar.

Andrei Nomerotski
Jesus College
University of Oxford

1

Unexpected Encounter

Misha left the main building of Moscow State University. Today the University was open to the senior grade high school pupils, and they were welcome to visit the lecture rooms and laboratories. Misha was dazzled and confused by the endless number of instruments, machines and facilities.

Outside, Spring was in full swing. It was one of those days in May when the land is flooded with bright sunshine and the young greens sparkle with emerald light.

Misha sat on a bench trying to clear his head. A young girl was passing by. She was beautiful — in a glance he took in her large dark eyes, long lashes, and lovely face. He followed her with his eyes thinking of making her acquaintance. It would be great, for example, if some hooligan tried to attack her and he, Misha, would come to her rescue. And the girl would give him a smile...

Suddenly Misha felt someone looking at him. Only now did he notice that he was sitting next to a gentleman in his fifties. Slim and tidy, he was staring at Misha with smiling eyes.

Misha flushed. He didn't like old people: they were usually very demanding and liked to interfere in everyone else's business.

"Young man, please do not look at an old and distinguished academic as if he's a dog," uttered the gentleman kindly, but firmly, rapping out the words. "When I was your age I used to think that all people over forty were stupid old fellows who were out of their minds

1

and needed to be retired. As you grow older you start viewing age differently. In England, for example, people over 60 are considered to be middle-aged. Well, I have not introduced myself yet," added the stranger. "I am Professor Leonid Andreevich Petrov. How should I address you?"

"Misha," mumbled Misha, very confused by meeting a famous scientist and by his familiar yet respectful manner.

"Misha is a very good name," said Leonid Andreevich. "Well, Misha, I heard you asking questions during the tour of the University. I liked your questions. Curiosity is very important in science, and so is the ability to ask questions, even if they are still not very sophisticated ones. Very often a fresh, inexperienced glance grasps more than the persistent and patient study of a professor who can't see beyond the end of his own nose.

"Well, Misha, I am busy with an investigation where a keen, inquisitive and direct mind is required, and you seem to possess all these qualities. Therefore, I invite you to help me. The summer vacation has already started, so you may have some spare time. If you're interested, please come tomorrow at ten o'clock in the morning to this address." with that the Professor gave Misha a business card with his office address. "You'll need to bring your ID card so that you can enter the building."

"All right," said Misha, and without adding anything else he shook the hand of the smiling Professor.

Next morning, at ten minutes to ten, Misha was standing at the entrance to a small building with an engraved plate: 'Institute of Systems Studies of the Academy of Sciences.' Misha gingerly entered the lobby and joined the queue of people waiting to be allowed in by the guard. Everyone showed him their passport and the guard, after a cursory glance at the name, took his time scrutinising a small piece of paper with something written on it. Now it was his turn: after a swift glance at his ID card, the guard carefully studied his pass as if trying to find some mistake, then handed it to Misha and said: 'Room 201, second floor.'

The door of room 201 bore a nameplate: "Professor Leonid Andreevich Petrov." Misha knocked at the door and opened it, but instead of the Professor he saw a young woman at a desk. The woman was pretty, but had an austere and discontented look about her. Misha was immediately reminded of his English teacher.

"Good morning, I was invited by Leonid Andreevich," said Misha and gave his name. Suddenly the face of the young woman — and Misha understood that she was the Professor's secretary — changed completely, expressing a warm welcome.

"Leonid Andreevich is expecting you." She pointed at a huge oak door.

Misha opened the door and saw Leonid Andreevich smiling at him with a telephone in his hand.

"Here you are, professor," joked Leonid Andreevich interrupting his conversation. "Take a seat while I finish my call."

While Leonid Andreevich was talking, Misha examined his room. It was a rectangular room with two tables — a desk which was occupied by Leonid Andreevich and a large table with chairs for conferences. Opposite the conference table there was a blackboard and some posters. One of the posters, beautifully made, read: 'Study of a pulsar' and below there was some kind of a diagram.

"Hello, Misha," Leonid Andreevich said merrily. He took a seat opposite Misha and said: "Yes, young man" — as if inviting him to start a conversation, and, quite unexpectedly, Misha asked "What is a pulsar?"

"That is a good question. Let us start with it."

And he told Misha the following.

2

What is a Pulsar?

In 1967 a powerful radiotelescope was built in Cambridge to study the interplanetary scintillation of radiowaves. Radiowave scintillations themselves were discovered somewhat earlier in 1964. Let us suppose that there is a radiowave source located very far from the Earth. The radiowaves travel long distances before they reach the Earth's surface, and on the way they pass through clouds of rarefied plasma in interplanetary space. These diffract the waves, and so scintillations occur on these clouds. The study of interplanetary scintillations allows us to obtain very important data on the angular dimensions of a radiosource.

Usually, when speaking of a radiotelescope, we imagine a huge parabolic mirror which, like a gigantic eye, is directed into the depths of space. However, the Cambridge telescope, or, to be more precise, the telescope of the Mullard Radio Astronomy Observatory, looked completely different. This huge structure resembled the eye of a fly or a dragon-fly, and consisted of more than two thousand dipole antennae. They filled a rectangular area of about eighteen thousand square meters. All the dipoles were precisely phased, so that they formed a 'compound eye'. By changing the phases of the dipoles, the 'direction' of the antennae could be changed also, i.e. the 'compound eye' could be rotated.

Routine operation began in July, 1967. The telescope permitted simultaneous observation of several different sectors of the sky. At the end of August, Miss Bell, whilst checking the recorded data, found a very peculiar dot. It looked like a scintillating source which flickered at about midnight when interplanetary scintillations were weak. As the amount of data grew, it became evident that the source was located far beyond the Solar System. The possibility of radiodisturbances from a far space object was excluded.

Further observations showed that this radiation was emitted in periodic pulses. You can imagine the excitement of Professor A. Hewish, Miss Bell and their colleagues, who encountered this new phenomenon. Estimates of each pulse duration showed that they were no longer than twenty milliseconds. This proved that the source was of planetary dimensions; the possibility that the signals came from an alien civilization couldn't be ruled out. The observations revealed that the pulses had an amazing regularity. Their repetition frequency was precise to better than 1 part in 10^7.

The search for similar objects very soon resulted in the discovery of three other pulsing objects. They were named pulsars. This was reported in 1968 and, starting from February, 1968, all the major radiotelescopes began to study these new space objects. For his role in the discovery of pulsars, Professor Hewish was awarded the Nobel Prize. Later many other pulsars were discovered. They all have an exceptionally precise repetition of pulses, with periods that range from several milliseconds to several seconds.

The natural explanation of a pulsar's precise repetition is that it is related to a star's rotation. In this case the star should be very compact. The existence of such stars had already been theoretically predicted, and they were called neutron stars. The substance of neutron stars resembles that of the nuclei of heavy atoms.

According to theory, neutron stars can be formed at the end of the evolution of massive stars. After the nuclear fuel of the star 'burns out', i.e. the majority of nuclei are Fe nuclei, the central part of the star gets compressed. The density grows and at a certain moment

Fe nuclei, which are located close to each other, disintegrate, which is very energetically favorable. The star's core collapses inwards at high velocity. At the same time, the star's shell is ejected: a supernova has occurred.

After such a flash, the core of the massive star implodes to a 'tiny' neutron star (its radius is only several kilometers), around which a swiftly expanding shell forms. As it moves away from the star, this shell transforms into a nebula. The most well-known nebula of this type is the Crab Nebula, and just inside this nebula one of the pulsars was discovered, or, to be more precise, a short-period pulsar, i.e. a relatively young neutron star.

Why does such a star emit radiowaves? It turns out that a neutron star is magnetized. Like the Earth, it has a dipole magnetic field. The only difference is that it is a hundred billion times stronger than the magnetic field of the Earth. As it happens, the superstrong magnetic field of a neutron star may drastically change the properties of matter. Additionally, this gigantic magnet, rotating very quickly, significantly excites the surrounding plasma. This very excited plasma becomes a radiotransmitter which generates radiowaves. The period of repetition of radiopulses is exactly equal to the period of the neutron star.

Simply speaking, a pulsar is a unique object in Nature!

3

Mischievous Ball

"And now we shall go and see what a pulsar is," said Leonid
Andreevich.

"What do you mean?" inquired Misha.

"You will find out very soon," answered Leonid Andreevich, and
he invited Misha to follow him.

They left the Professor's room and went to the very end of the
corridor. Leonid Andreevich stopped near a large door and pressed
some buttons on the coded lock. The door slowly moved aside and
they stepped in. Misha never expected to find such a room in the
building: they came into a huge round hall with a hemispherical
dome which was evenly illuminated by daylight. To tell the truth,
Misha thought that they were outside the building, and only after
several moments did he understand that they were still indoors. The
hall reminded him of the Planetarium, the only difference being that
there were no chairs on the periphery. In the centre of the hall there
was a large round table instead of the projector that showed the sky
and stars. Around the table there were several chairs which looked
like dentist's chairs: they could be turned and reclined so that it was
very easy to look upwards. Opposite one of the chairs, a large display
was installed and it had a computer keyboard beside it. Numerous
different things were scattered all over the table. Among them Misha
noticed some large glass plates and a box with balls which looked

like ping-pong balls. Leonid Andreevich took the chair opposite the display and invited Misha to take a seat next to him. The display showed one line of text, 'Earth, day, cloudy' and some numbers.

Leonid Andreevich pressed one of the keys. The display went blank and he typed 'Pulsar' and then some numbers and pressed the same key. Suddenly the hall was transformed. The floor around the table was emitting a bright blue light and the dome had become a black night sky with myriad shining stars.

"We are on a pulsar already", said Leonid Andreevich. "Now we shall do some experiments, and, to make it easier to see, I shall switch on the Sun, which, of course, you would not be able to find on a pulsar. For our comfort I had better hide it behind some clouds."

Leonid Andreevich keyed in some numbers and a bright daylight filled the hall again.

"We shall start with the simplest experiments," said Leonid Andreevich. He took one of the balls from the box, lifted it above the glass plate on the table and let it drop. The ball started to bounce, the height of each bounce slowly decreasing until, after some time, the ball came to rest on the plate.

"This is a very good ball, it bounces for quite a long time," explained Leonid Andreevich.

"If the ball did not lose energy at each bounce, it would continue bouncing forever. Do you think that if we observed these bounces of smaller and smaller amplitude through a microscope, we'd see that this ball bounced forever too?" asked Leonid Andreevich.

"I don't know," answered Misha, puzzled.

"Well, let us think. Let us assume that at any stroke the ball loses a small fraction of its momentum. Let us denote this fraction by ϵ. In other words, if the ball's velocity before a stroke is v, then after a stroke it will be equal to $v - \epsilon v$. We assume that there is no air resistance. Then, before the second stroke the velocity will be the same, i.e. $(1 - \varepsilon)v$, and after the second stroke $(1 - \varepsilon)(1 - \varepsilon)v$. So, after the nth-stroke, it is equal to $(1 - \varepsilon)^n v$. If the ball is initially thrown upwards from the surface with velocity v_0, then it will drop back after $t_0 = 2v_0/g$, where g is the free-fall acceleration. The time of the subsequent bounce will be $t_1 = 2v_1/g = 2(1 - \varepsilon)v_0/g$, and, consequently, $t_n = 2(1 - \varepsilon)^n v_0/g$. Thus, the total time of the ball's bounces is equal to

$$t = \sum 2(1 - \varepsilon)^n v_0/g$$

where the $\sum$ symbol denotes the sum over all n, beginning from $n = 0$ and ending at infinity.

"How can we estimate this sum? It is very simple. Let us say that $x = (1 - \varepsilon)$. We wish to find the value of $A = \sum_{n=0}^{\infty} x^n$. Multiplying A by $1 - x$, let us write down the result in two lines, one above the other.

$$(1 - x)A = 1 + x + x^2 + \cdots$$
$$-x - x^2 - \cdots$$

We see that the second line added to the first line gives unity. It follows that $(1 - x)A = 1$, i.e. $A = 1/(1 - x)$. Thus, the total time is

$$t = \frac{2v_0}{g} \frac{1}{1 - x} = \frac{2v_0}{g\varepsilon}$$

We have certainly treated infinite series rather freely, but actually the trick of summation of infinite series that we have used accords with the results of more sophisticated mathematics.

"Thus, we have found out that the ball will bounce only within a limited time t, after which it stops completely.

"By the way, this postulate solves one of Zeno's famous 'aporia'. In his time the Greek philosopher Zeno puzzled his compatriots by unusual statements which he called 'aporia'. One of them was, 'Achilles can never catch up with a tortoise'. Here is how he was thinking:

"Let Achilles start running after the tortoise at some given time. When he reaches the place from which the tortoise started, it will have moved ahead a little bit. Achilles must reach this new point to catch the tortoise, but by the time he reaches it the tortoise will have moved farther away, and so on. It seems that Achilles will never catch the tortoise.

"The solution is that although it takes an infinite number of these steps for Achilles to catch the tortoise, the total time is definitely finite, so Achilles will catch up with the tortoise rather quickly. However, as Frenchmen say, *revenons à nos moutons* ...

"Up until now we have been throwing the ball on Earth. Now let us go to the pulsar," and the Professor once more played with the computer keyboard, then he took the ball, lifted it over the glass plate, and released it. To Misha's amazement the ball started to fall at an angle to the table, then hit the plate, and bounced back along the same inclined line and continued bouncing as if it were tied to an invisible thread. Finally, it came to rest. The Professor repeated the trick and said: "And now I am going to explain what just happened.

"The magnetic field of a pulsar is very strong, so it even affects the movements of regular, non-magnetic objects. The superhigh magnetic field produces a mass anisotropy: the mass along the magnetic field direction $m_\|$ maintains its value, i.e. $m_\| = m$, whereas across the magnetic field the mass $m_\perp$ considerably exceeds m. In other words, Newton's law along the magnetic field keeps its original expression:

$$ma_\| = F_\|$$

where $a_\|$ is the acceleration along the magnetic field, and $F_\|$ is the force component along the magnetic field.

"Across the magnetic field Newton's law is:

$$m_\perp a_\perp = F_\perp$$

It turns out that $m_\perp = m(1 + b)^2$ where b is equal to the magnetic field, B, divided by a certain constant, B_0, of the order of 10^9 Gauss[*]. The magnetic field on the pulsar can reach and even exceed 10^{12} Gauss, so $m_\perp$ can be significantly larger than m, i.e. the mass anisotropy can be very high.

"Let us assume that on our pulsar $b^2 \gg 1$, since high anisotropy is typical of a pulsar, and yet is very unusual for us.

---

[*] 1 Tesla = 10 000 Gauss. Throughout the text magnetic field strengths will be given in Gauss. However, the accompanying formulae require B to be expressed in terms of Tesla.

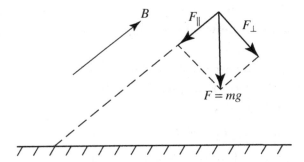

Fig. 3.1   The gravitational force $\boldsymbol{F} = m\boldsymbol{g}$ can be decomposed into two components: the longitudinal component $\boldsymbol{F}_{\parallel}$ is directed along the magnetic field $\boldsymbol{B}$, and the transverse one $\boldsymbol{F}_{\perp}$ is directed perpendicular to the field.

"Imagine that the magnetic field is inclined to the horizontal plane as shown in Fig. 3.1. The gravitational force, $F = mg$, is, of course, directed straight down. Note that this expression contains m, but not $m_{\perp}$. This is because the ball does not actually become heavier: the use of a $m_{\perp}$ value is simply a convention that accounts for the effects of the magnetic field.

"This force can be split into two components: along and across the magnetic field, i.e. $F_{\parallel}$ and $F_{\perp}$ respectively. Then Newton's law can be applied:

$$m_{\parallel}a_{\parallel} = F_{\parallel}, \quad m_{\perp}a_{\perp} = F_{\perp}$$

As $m_{\perp} \gg m_{\parallel}$, the acceleration perpendicular to the magnetic field, $a_{\perp}$, turns out to be much less than the longitudinal acceleration. That is why the ball 'slides' almost along the magnetic field. We can easily find the distance, S, between where the ball lands and the intersection of the magnetic field with the horizontal plane. Indeed, if the initial velocity is zero, then the path length L along the field is given by the relation $L = F_{\parallel}t^2/2m_{\parallel}$, where t is the time of the fall. The transverse deviation is equal to $S = F_{\perp}t^2/2m_{\perp}$ so that:

$$\frac{S}{L} = \frac{F_{\perp}}{F_{\parallel}}\frac{m_{\parallel}}{m_{\perp}}$$

This relation is less than unity if the ratio of $F_\perp$ to $F_\parallel$ is not too high, i.e. if the angle between the magnetic field and the plane isn't too small.

"The values of the acceleration and velocity along and across the field differ greatly. Meanwhile, according to classical mechanics, the momenta $p_\perp$ and $p_\parallel$, are:

$$p_\parallel = m_\parallel v_\parallel = F_\parallel t, \quad p_\perp = m_\perp v_\perp = F_\perp t$$

It is easy to see that the total momentum lies in the same direction as the force due to gravity, i.e. vertically. Therefore, the momenta are not affected significantly by the mass anisotropy.

"Now let us try to understand what happens when the ball hits the horizontal plane. If the collision is elastic, the vertical component of momentum will change sign. Since there is no horizontal momentum component, this means that the velocities $v_\parallel$ and $v_\perp$ will change sign too: the ball will run backwards along the initial trajectory. The amplitudes of subsequent bounces will decrease due to small energy losses until the ball comes to rest.

"Now, let us consider another experiment. Let's throw the ball with a slight horizontal push. Now the collision point will drift along the plane between bounces since the horizontal component of the ball's momentum is nonzero. The resulting trajectory will look like that in Fig. 3.2a. It differs drastically from the analogous trajectory on Earth (Fig. 3.2b).

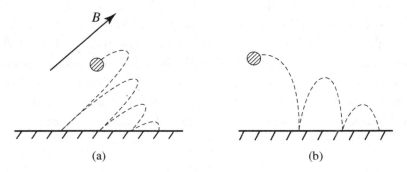

(a)   (b)

Fig. 3.2 A ball in a super-strong magnetic field bounces in a dramatically different way (a) to normal bouncing (b).

"First of all, note that a horizontal push to a ball on a pulsar causes not only horizontal, but also vertical displacement (this is impossible for an object with isotropic mass!). Besides that, it bounces radically differently from how it would on Earth. If it were on Earth, i.e. if the mass were isotropic, the angles on each side of a bounce would be equal because the horizontal momentum would be conserved (for elastic impacts), so the ball would bounce in the familiar manner shown in Fig. 3.2b. In the case of an anisotropic mass, the trajectory looks as is shown in Fig. 3.2a. From bounce to bounce the ball gradually shifts along the horizontal plane."

"And what will happen if instead of a ball we take a bead on a stretched thread?" asked Misha.

"Oh, this is an excellent question," answered the Professor. "I can even show you what will happen."

The Professor rummaged in some boxes on the table, took out a string from one of them, pulled it through a big bead and then fastened the string at both ends so that it hung vertically. He lifted the bead and then dropped it. The bead started to slide along the string like in a slow motion film: it descended with constant acceleration, but at a much slower rate than in free fall.

"Let's try to understand the reason behind this," said L.A.

"Let the vector B of the magnetic field form an angle α with the string. A bead with a strongly anisotropic mass would tend to fall along the magnetic field line, but the string prevents it: a force N is produced by the string and is directed perpendicularly to the string. As a result, the bead will fall along the string with an acceleration a. It's a simple problem in mechanics to find this acceleration if the forces are known. However, in our case we only know one of the forces — the force due to gravity. As for the force N, it is unknown, and we cannot ignore it.

"In solving similar problems, the dynamic task is often artificially reduced to a static one. For example, Newton's second law for an isotropic mass can be written as:

$$\boldsymbol{F} + \boldsymbol{F}_i = 0$$

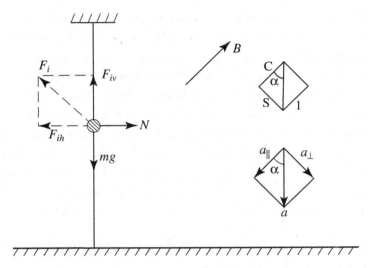

Fig. 3.3   When the motion of the bead is constrained by a string, the bead experiences a normal reaction force $\boldsymbol{N}$ in addition to the usual gravitational force $m\boldsymbol{g}$. In this case, the inertial force $\boldsymbol{F}_i$ has two components: a vertical one F_{iv}, and a horizontal one F_{ih}.

where $\boldsymbol{F}$ is the driving force and $\boldsymbol{F}_i$ is an artificially introduced inertial force which is equal to $\boldsymbol{F}_i = -m\boldsymbol{a}$. The net force, i.e. the sum of the forces $\boldsymbol{F}$ and $\boldsymbol{F}_i$, obeys the equilibrium condition. Let's use the same approach for our problem. If the acceleration of the bead is equal to a and is directed vertically down, it can be decomposed into two components: along and across the magnetic field (see Fig. 3.3).

Now we can find the components of the inertial force:

$$F_{i\parallel} = -m_{\parallel}a_{\parallel}, \quad F_{i\perp} = -m_{\perp}a_{\perp}$$

$\boldsymbol{F}_i$ is equal to the sum of the forces $\boldsymbol{F} = m\boldsymbol{g}$ and $\boldsymbol{N}$. Splitting $\boldsymbol{F}_i$ as shown in Fig. 3.3 into horizontal and vertical components, we find that the horizontal component balances $\boldsymbol{N}$, whilst the vertical component balances $m\boldsymbol{g}$. Therefore, if we are able to express F_{iv} in terms of the acceleration a, then, by setting this equal to $m\boldsymbol{g}$, we will find the acceleration. If you know geometry this isn't a difficult problem, and it is even simpler if you know trigonometry.

"To further simplify our considerations, let's introduce an auxiliary rectangle with its sides along and across the magnetic field and the diagonal equal to unity. The sides of this rectangle will be c and s, respectively (those who know trigonometry will immediately recognize that $c = \cos \alpha$, $s = \sin \alpha$, where α is the angle between the string and the magnetic field, so we will also use this more advanced notation).

"From the similarity of the rectangles — for a and for the auxiliary rectangle — we obtain:

$$a_{\parallel} = ac = a \cos \alpha, \quad a_{\perp} = as = a \sin \alpha$$

The inertial force $\boldsymbol{F}_i = -m_{\parallel} \boldsymbol{a}_{\parallel} - m_{\perp} \boldsymbol{a}_{\perp}$. The vertical component F_{iv} of this force is equal to the sum of the vertical components of each term. From the law of similar triangles we easily find:

$$F_{iv} = -m_{\parallel} a_{\parallel} \cos \alpha - m_{\perp} a_{\perp} \sin \alpha$$

Substituting in the values of $a_{\parallel}$ and $a_{\perp}$ we obtain: $F_{iv} = -(m_{\parallel} \cos^2 \alpha + m_{\perp} \sin^2 \alpha) a$. However, we derived this on the condition that $F_{iv} + mg = 0$. Therefore, we have:

$$(m_{\parallel} \cos^2 \alpha + m_{\perp} \sin^2 \alpha) a = mg$$

It looks as if the bead is getting heavier: its effective inertial mass is equal to $m_* = m_{\parallel} \cos^2 \alpha + m_{\perp} \sin^2 \alpha$. In the case of a very strong anisotropy, $m_{\perp} \gg m_{\parallel}$, the mass m_* appears to be much larger than $m_{\parallel}$, if the angle α isn't too small. The bead, therefore, falls along the string with an acceleration significantly lower than g: the fall appears to be moderated in time. Still, we have to make sure that the law of conservation of energy is satisfied. The velocity of the bead at time t is equal to $v = at$, where t is measured from the beginning of the fall with zero initial velocity. The distance S fallen through in time t is equal to $at^2/2$. So, if an object falls from height h, $S = h = at^2/2$. h may be expressed in terms of the velocity: $h = v^2/2a$. Since $a = gm/m_*$, we obtain: $\frac{1}{2} m_* v^2 = mgh$. So we see that the kinetic energy at the end of the path is equal to the potential energy at height h.

"We are now ready to describe the movement of a bead along a string inclined at an arbitrary angle to the horizontal. First, note

that our treatment of forces for a vertical string can also be used for
the momentum components, i.e. if v is the particle velocity along the
string, the longitudinal and transverse velocity components (along
and across the magnetic field) are:

$$v_\| = v \cos \alpha, \quad v_\perp = v \sin \alpha$$

Let us consider the momentum of a particle $\boldsymbol{p} = m_\| \boldsymbol{v}_\| + m_\perp \boldsymbol{v}_\perp$. The
momentum component along the string is equal to:

$$p = (m_\| \cos^2 \alpha + m_\perp \sin^2 \alpha)v = m_* v$$

As we know, acceleration is the rate of change of velocity. Thus,
Newton's equation for motion along the string can be presented as a
balance between the rate of change of momentum and the external
force. The m_* value depends only on the orientation of the string
with respect to the magnetic field, i.e. it can be used for the inclined
string, minding the angle between the string and the magnetic field.
The gravity force component along the inclined string will be less,
of course, since it is well-known from the motion along the inclined
plane. Further deliberations are simple enough for you to go on with
them independently.

"Now we shall perform another experiment," said L.A.

"Let us take a ball and bounce it not on the horizontal plane,
but on the plane perpendicular to the magnetic field. From Fig. 3.4
we can see that the ball, bouncing, slides along the plane. This case
is simple enough, since the motions along and across the magnetic
field are not connected with each other: the ball bounces along the
magnetic field because the mass $m_\|$ is subjected to the component of
the force of gravity along the field; it slides along the plane because
the transverse mass $m_\perp$ is subjected to the transverse gravity com-
ponent. As $m_\| \ll m_\perp$, the ball bounces rather frequently, but moves
slowly across the plate."

"That wasn't hard to understand," said Misha, who followed
L.A.'s deliberations only with some difficulty.

"Now let us do the following," said L.A.

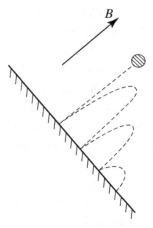

Fig. 3.4 Ball bouncing in the case when its mass is highly anisotropic due to the presence of a superstrong magnetic field **B**.

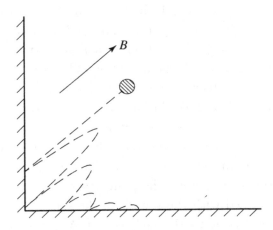

Fig. 3.5 Ball bouncing near a corner in the case of a very high mass anisotropy.

"Let us take a couple of plates and place them one vertically, the other horizontally at right angles to each other. Let the ball first fall on the vertical plate." L.A. released the ball, and it fell along the force line in the direction of the vertical plate. Then, bouncing, the ball slipped down along this plate and rolled along the horizontal plane (Fig. 3.5).

"The ball gains horizontal momentum when it collides with the vertical plate. Part of the kinetic energy of the longitudinal motion is transferred into transverse (with respect to the magnetic field direction) kinetic energy. Thus, even in the ideal case of elastic collisions, the amplitude of the bounces along the magnetic field decreases gradually from stroke to stroke. At last, the ball descends along the wall so that its average fall looks like the bead motion along the vertical string. The ball possesses fairly large horizontal momentum when it reaches the horizontal plate. The ball starts bouncing as if it were thrown with an initial horizontal push.

"Well, you have been introduced to the simplest cases of motion with a highly anisotropic mass. This is the way that objects should move in a pulsar's superstrong magnetic field."

"Are we really on a pulsar?" Misha seemed to be surprised.

"Well, no." L.A. smiled with the boy. "We are in a lab which can simulate the conditions on a pulsar. This hall is designed to study adaptation, i.e. the accommodation of a man to unusual external conditions. The walls and the dome of the hall are made of numerous LCD screens. These displays receive signals from a very powerful computer capable of reproducing a consistent picture from a great number of images." "Here is an example."

L.A. keyed in some command and Misha almost cried out with surprise: he found himself on the beach of a tropical island. The surf gently licked the sandy shore, the palm trees swayed in the breeze, and the air was filled with the sounds of the jungle.

"Now let us return to Russia." said L.A., and in a couple of moments they found themselves on a lawn somewhere in the suburbs of Moscow.

"This round table is unusual too," said L.A. "Inside it has a complicated system of coils through which currents are passed to produce alternating magnetic fields. The computer automatically controls these fields so that the balls, which are equipped with special internal structures, are capable of rather complicated motions. In particular, it is very easy to simulate the motion of objects with anisotropic mass. We simulate only a portion of the conditions on the pulsar. This is why I speak of a 'trip to the pulsar'. Later, we will introduce more variations into our experiments. Your task is simply to carefully observe everything, ask questions and try to understand what is going on.

"Well, that is enough for today," added L.A. "Come back in a week's time. Meanwhile, to keep you busy, I would like you to work through this at home," — and he gave Misha several pages of printed material. The first page was titled: Problem No. 1 and it contained text with equations and figures (it is given in the Appendix).

Misha said goodbye and hurried home.

4

Misha and Sasha

Having spent some time at home, Misha decided to visit his friend Sasha. Sasha seemed very smart to him as he was a third year student at Moscow State University's Physics Department. Sasha was always ready to answer any question in physics and mathematics and Misha thought that he knew almost everything about physics. While speaking about his talk with the Professor, Misha briefly mentioned mass anisotropy in a strong magnetic field. Sasha's reaction was very abrupt:

"You must have mixed things up. There is no such thing as mass anisotropy. I have just studied the physics of magnetic phenomena and I know this field inside out!

"Almost all matter belongs to one of three classes of magnets — diamagnets, paramagnets, and ferromagnets. Diamagnets and paramagnets do not possess an intrinsic magnetic moment, they are only magnetized in the presence of a magnetic field. In diamagnets, an applied magnetic field gives rise to a magnetic moment which is directed against the field, whereas in paramagnets the induced magnetic moment is directed along the field. Paramagnets seem to consist of a multitude of elementary magnets which, in the absence of a magnetic field, are randomly oriented so that their total magnetic moment is equal to zero.

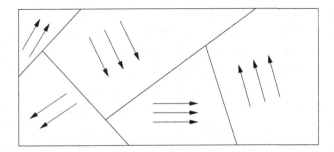

Fig. 4.1 Ferromagnets look like a composition of magnetic domains: each of them is magnetized up to saturation.

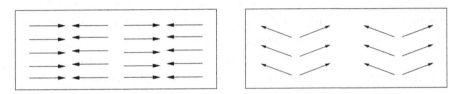

Fig. 4.2 Antiferromagnets (a) and weak ferromagnets (b) have two equal magnetic sub-lattices which are directed either in exactly opposite directions (antiferromagnets) or almost opposite each other (weak ferromagnets).

"In the presence of a magnetic field they begin to align with the field, and a total magnetic moment is produced: the higher the magnetic field, the stronger the moment. The elementary magnetic moment of a single atom is sometimes called a 'spin'. To be more precise, the spin is the angular momentum of the atom's electron shell, but the related magnetic moment is also called a spin. By the way, besides the electron shell, the atom's nuclei also has its own magnetic moment, but it is much smaller than the electron shell moment.

"Diamagnetic materials have no intrinsic magnetisation; a magnetic field 'brings' the magnetic moment on them by affecting the orbital motion of their atoms' electrons.

"The magnetization of diamagnets and paramagnets is usually low. That is why ferromagnets are more widely used in technology. Iron is a ferromagnet, the word is derived from the Latin word for iron, 'ferrum'. Ferromagnets become magnetized in comparatively low magnetic fields. For example, if you leave a rod made of soft iron oriented from north to south overnight, you will find it magnetized by morning. You can test this with the help of a compass.

"Ferromagnets are magnetized so easily because their atoms initially possess a high magnetic moment. Each piece of a ferromagnet consists of a number of small strongly magnetized areas — small magnets, so to speak. Each area is called a magnetic 'domain'. Usually, the magnetic moments of these domains are randomly oriented so that the average magnetic moment equals zero.

"As soon as an external magnetic field is applied, the domains begin to align. This alignment produces a net magnetic moment. As the field increases, all the domains line up along the field and the ferromagnet becomes fully magnetized.

"Besides these three simple classes of magnets (i.e. diamagnets, paramagnets, and ferromagnets), there are substances with more complex magnetic properties. These materials have two or more types of magnetic atoms, so there can be two or even more magnetic lattices in a solid crystal. If two sub-lattices have magnetic moments of equal magnitude, but opposite orientation, the material is called anti-ferromagnetic. If the antiferromagnetic sub-lattices are not exactly similar, a net magnetic moment appears. Such a crystal is called ferrimagnetic. In some materials, which resemble antiferromagnets, the crystalline symmetry means that the two sub-lattices are not precisely anti-parallel — they appear to be slightly inclined with respect to each other. This sort of crystal is called a weak ferromagnet.

"There are other more complex magnetic structures. For instance, a certain layer of magnetic moments in the magnetic sub-lattice may not repeat precisely in a neighboring layer, but be slightly turned in comparison to the previous layer.

"Finally, the most complex magnetic material is the 'spin glass'. In the magnetic spin glass, every elementary magnet interacts with its spin environment in a very complicated way. With respect to certain spins it tries to settle in a parallel way; whereas in respect to others, it tries to settle in an anti-parallel way. This results in a state with randomly distributed spin directions due to their interaction, not simply because of thermal motion. It turns out that the spin glass is the simplest example of a neural net, i.e. a primitive analog of the neural network of the brain."

"Why are the magnetic structures so different?" asked Misha. "Does it mean that individual magnetic moments interact just like the magnetic hand of a compass, i.e. their even polarities are pushed away whereas their odd polarities attract each other?"

"No," answered Sasha. "The atoms inside a crystal possess magnetic moments. They are guided not by classical, but by quantum mechanics. According to quantum mechanics, a turn of each spin results in a deformation of the electron density of a crystal, i.e. it is accompanied by an additional change in the collective energy of all the electrons. The interaction of the spins becomes more complex due to the 'stirring' of a crystal's electron component.

"Thus, the magnetic properties of matter may be very diverse. However, they show nothing similar to mass anisotropy. Therefore, Misha, you either misunderstood something, or you need to ask the Professor how a magnetic field may lead to mass anisotropy.

"Nevertheless, mass anisotropy proper is a very interesting physical phenomenon: I have even realized how to model it. I have a toy at home — a small car with an inertial motor. The motor is a fly-wheel that is driven by means of a gearbox. If you place the car on the floor and start pushing it, the fly-wheel gains momentum and then slowly releases energy sufficient for the car to move. It is also possible first to accelerate and then to restrain the toy, speeding up and slowing down the fly-wheel, respectively. In doing so you develop a feeling that you are handling a rather massive object which is very difficult to accelerate or slow down. Moving this toy in a straight line inevitably drives the fly-wheel up to speed, whereas its lateral motion is as for any other object — so we have a model of an object with mass anisotropy. To get this effect, one can replace smooth wheels by toothed ones and make a corrugated floor. Then, while pushing the toy across the ribbed floor, its wheels will fit into the grooves and the fly-wheel will be forcefully driven, whereas along the ribs at low friction the toy will move freely with its normal mass. Therefore, we obtain an object with mass anisotropy, the direction of which is preset externally."

5

Playing Billiards

In a week Misha was back in the same hall.

"Today we will play billiards," said L.A. "Not ordinary billiards, but billiards on a pulsar. All the balls have a strongly anisotropic mass, i.e. their transverse mass $m_\perp$ significantly exceeds their longitudinal mass $m_\parallel$. For convenience, we will direct the magnetic field horizontally so that the balls will appear light in the longitudinal direction (along the table) and heavy in the transverse one (across the table).

"First we must learn how to hit the balls with a cue. If you hit the ball at an arbitrary angle it will move almost longitudinally. Indeed, if you apply an arbitrary momentum $\boldsymbol{p}$ to the ball, then resolving it into longitudinal and transverse components gives:

$$\boldsymbol{p} = \boldsymbol{p}_\parallel + \boldsymbol{p}_\perp = m_\parallel \boldsymbol{v}_\parallel + m_\perp \boldsymbol{v}_\perp$$

As $m_\perp \gg m_\parallel$, the transverse speed will be low compared to the longitudinal one, i.e. $v_\perp \sim v_\parallel m_\parallel / m_\perp$. To direct the ball in an arbitrary direction, i.e. in order to have $v_\perp \sim v_\parallel$, one has to hit the ball in such a way that the longitudinal momentum is small, i.e. $p_\parallel \sim m_\parallel p_\perp / m_\perp \ll p_\perp$. In other words, the cue should be aimed within a small range of angles $\sim m_\parallel / m_\perp$ from the transverse direction. This is not easy, but it is possible.

"Now let us find out what happens when a moving ball hits a ball at rest. We begin with regular balls on Earth. After a direct hit, the

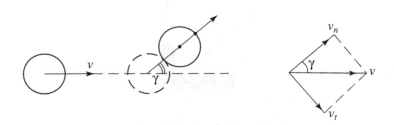

Fig. 5.1   Decomposition of the relative ball velocity v into tangential, v_t, and normal, v_n, components.

hitting ball stops and the one that was at rest moves with the speed of the hitting ball. This is known from the the conservation laws of energy and momentum — only when the energy and momentum are fully transferred are their net values conserved. Now let's consider an indirect hit (Fig. 5.1). Suppose that at impact the line connecting the centres of the two balls makes angle γ with the velocity of the hitting ball.

"First let us decompose the velocity of the hitting ball into two components: the normal to the tangent plane (at the collision point) v_n and the tangent component v_t. With respect to the normal component the impact looks like a direct one: the normal momentum of the first ball disappears while the second ball acquires it. The tangent momentum (i.e. the product of the ball's mass and v_t), cannot be transferred to the second ball as there are no tangent forces. So after the collision the balls will move away from each other at a 90° angle: the first ball will move with velocity v_t, and the second will move with velocity v_n.

"Now let us return to billiards on the pulsar. We shall start off with a direct hit: the first ball stops after the collision, and the second one moves with momentum equal to that of the first ball. Geometrically, the direct hit looks as expected only in the case when the balls move precisely across or along the magnetic field. An unusual peculiarity arises in the case of an arbitrary direction of the balls' motion.

"The reason is that, if the momentum of the striking ball is to be completely transferred to the stationary one, the tangent plane

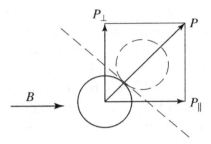

Fig. 5.2 The momentum P of the ball when it experiences a frontal hit with another ball (shown by the dotted line) can be split into a longitudinal and a transverse component.

must be perpendicular to the momentum at the collision point (see Fig. 5.2). Momentum, however, is connected with velocity via the relation:

$$p = m_{\|}v_{\|} + m_{\perp}v_{\perp}$$

Since $m_{\|} \ll m_{\perp}$, $v_{\perp}$ turns out to be much lower than $v_{\|}$, as long as $p_{\|}$ isn't very small. In other words, most direct hits result in the balls moving at small angles to the magnetic field. It can be said in a different way: if we send the first ball in a direction close to the magnetic field direction, then even after an indirect hit, the second ball will also move at a small angle to the magnetic field. (For the ball to move at a significant angle to the magnetic field, its transverse momentum has to be much higher than the longitudinal one, which is only possible with an almost entirely transverse hit).

"Now let's consider the case of an arbitrary collision. Again the momentum p can be split into normal p_n and tangential p_t components. In an elastic collision, the normal component is transferred to the second ball whereas the tangential one stays with the first ball. Therefore, after the collision, the momenta of the balls are once more perpendicular to each other, and the momentum of the second ball is directed from the point of collision to the centre of the second ball. From the point of view of momenta, the whole picture is the same as in billiards with isotropic balls. However, we observe speeds, not momenta. If we want to control the speed of the second ball, so that after the hit $v_{\parallel}$ and $v_{\perp}$ for the second ball are of the same order of magnitude, we need it to only gain a small momentum component along the magnetic field. To do so, it should be hit by the first ball in such a way that after the collision it gets a transverse momentum. Thus, the game requires the balls to be hit in a direction almost across the magnetic field (i.e., across the billiards table). This type of play is much more difficult than with isotropic balls so I've had to make the pockets much wider than on a regular table.

"Well, let us start," said L.A. inviting Misha to the table.

They played several games, until Misha had grown accustomed to the strange behavior of the balls.

"Now I will introduce you to a new game," said L.A. "which is simpler than billiards, but can only be played with balls of anisotropic mass. The system of ball movement control which we created allows us to simulate not just homogeneous magnetic fields, but also inhomogeneous fields of complex configurations. In particular, we can simulate the magnetic field of a cylindrical conductor with current flowing through it (Fig. 5.3). This magnetic field has an azimuthal direction, so that the magnetic field lines are concentric circles. The field is inversely proportional to the distance from the conductor. The transverse particle mass in a superhigh magnetic field with this

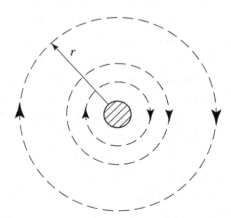

Fig. 5.3   Magnetic field lines of a straight conducting rod with current flowing through it.

geometry looks like $m_\perp = m_\parallel (1 + a^2/r^2)$, where r is the distance from the conductor, and a characterizes the radius where $m_\perp = 2m_\parallel$. Roughly speaking, when further away than a from the cylinder, the particle behaves as if it has isotropic mass, and when closer than $r = a$ the anisotropy begins to act; swiftly increasing as the particle moves nearer to the conductor.

"Now let's throw the ball through this superhigh magnetic field." L.A. lifted a small ball above the table on which a horizontal conducting rod was affixed, and let the ball go so that it passed close to the conductor. To Misha's surprise, the ball unexpectedly made a curve as if it were tied to the conductor by an invisible thread. Then it continued its route down along the familiar parabolic trajectory (Fig. 5.4).

"It may seem strange," said Leonid Andreevich, "but we just saw the 'free fall' of a body with anisotropic mass. Certainly, it is not free in the literal sense of the word because the lateral mass of the

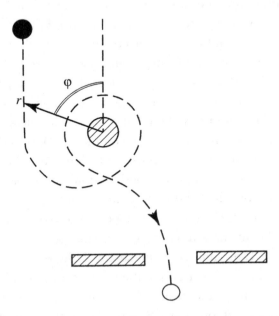

Fig. 5.4 Fall of an anisotropic mass ball through a region containing an azimuthal magnetic field.

body changes along the trajectory. Nevertheless no direct forces are applied to it.

"So, how can all of this be explained? It is convenient to consider the ball's motion in polar coordinates r, φ.

"Since we released the ball from a height it acquired a high speed. Consequently, in the vicinity of the conductor we can neglect the velocity change caused by gravity, i.e. we can assume the motion to be free. In the field under consideration the mass is conserved during azimuthal motion. Because of the symmetry, i.e. since there are no forces acting in the azimuthal direction, the angular momentum $\mathcal{M} = m_\parallel v_\varphi r$, where v_φ is the azimuthal velocity and is equal to ωr, where ω stands for the angular rotation frequency (the rate at which the angle φ changes), must be conserved. The smaller r is, the larger v_φ must be, i.e. the larger the centrifugal force acting in the r-direction.

"It can be said that the body is influenced by the force from a centrifugal potential. The radial movement is based on the fact that the body initially moves to the centre, but under the influence of the centrifugal potential recoils, reflects and flies back. In the case of a particle with isotropic mass, the azimuthal and radial movements combine such that the resultant is a straight trajectory. If, however, $m_\parallel \neq m_\perp$ and, moreover, $m_\perp$ increases as r decreases, the radial movement in the vicinity of the conductor slows down. As a result, the body starts to spiral towards the conductor. Due to the centrifugal potential, the ball stops its movement along the radius and then is pushed out again, leaving the superhigh magnetic field. Depending on the distance from the conductor from which the ball was dropped, i.e. on the 'impact parameter', the angle of the ball's flight from the region of significant mass anisotropy will change.

"The aim of the game is to make the ball penetrate a window located lower than the current conductor (see Fig. 5.4)."

Misha learnt the game easily. After the session he explained to his friends, and, first of all to Sasha, the unusual ways in which anisotropic mass bodies moved.

Sasha was astonished and could not understand what was meant. He had never come across such phenomena in any of his textbooks.

"Do you think I could be introduced to your Professor?" Sasha asked.

"He's a very kind person," said Misha, "I'll ask him next time."

6

Stubborn Fountains

"Which of two similar-sized bubbles submerged in water do you think will rise faster — an air bubble or a hydrogen one?" asked L.A. at their next meeting.

"The hydrogen bubble, of course" said Misha without thinking, "hydrogen is lighter than air."

"This time you have been too fast," said L.A. with a smile. "Let us think more thoroughly.

"According to Archimedes' principle, the upwards force on the bubble is $F_A = \rho g V$, where ρ is the density of the liquid and V the volume of the bubble. In this case, according to Newton's law we have:

$$ma = F_A - F_D$$

Here a is the bubble's acceleration, and F_D is the drag force experienced by the bubble while moving in the liquid. The drag increases with the speed, so the motion of the bubble is described as follows: at the moment of the bubble's take-off from the point of its formation (e.g. if it is found on the bottom of a tea-pot or if it is slowly expelled from a thin tube in an aquarium) its speed is close to zero, so the drag force is small and the bubble accelerates upwards. As the speed grows, F_D grows too until we have equilibrium between F_D and F_A, so the speed of the bubble becomes constant. F_D and, therefore, the speed of the steadily rising bubble depend on the liquid's viscosity: the same bubble in honey, for example, rises much more slowly than in water. Both F_D and the Archimedes force F_A, depend on the bubble's size such that small bubbles rise more slowly than large ones.

"Let's assume, for the sake of simplicity, that the viscosity is vanishingly small. Then which mass is m? It would seem that $m = \rho' V$ where ρ' is the density of the gas in the bubble, but this is not true: this assumption contradicts our routine observations. The density of air is a thousand times lower than that of water and, therefore, the air bubble in the water would be accelerated by a thousand times g, which is certainly incorrect. To understand what is meant by the mass m, let us consider how the bubble is rising. If the bubble moves with speed v, the liquid streams around the bubble with a velocity proportional to v (see Fig. 6.1). Therefore the Archimedes force, whilst lifting the bubble, also makes the liquid around the bubble move.

"For further deliberations it is convenient to use energy since energy is being conserved (if we also consider the kinetic energy transformed into heat due to the viscosity).

"It is obvious that the liquid which streams around the bubble has a given kinetic energy E which increases as the speed of the

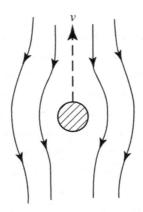

Fig. 6.1   The water stream around the gas bubble.

bubble increases. To find E, it is necessary to find the kinetic energy of selected small portions of the liquid and then to sum all these energies. However, the result is obvious in itself — since the volume of streaming liquid is proportional to the bubble volume V, and its speed is proportional to the bubble speed then:

$$E = \frac{1}{2} A \rho V v^2$$

Here A is some constant of proportionality. The kinetic energy of a body is equal to $mv^2/2$ where m is its mass. Therefore, as the mass in Newton's law we have to use the value

$$m = A\rho V$$

Here ρ is the density of the liquid. If desired, the mass $m' = \rho'V$ of the gas could be added to this, but since it is negligibly small compared to $A\rho V$ we can neglect it. Likewise, we can neglect the force of gravity acting on the gas so that, at this level of accuracy, there is no difference between the air and hydrogen bubbles. The value $m = A\rho V$ is usually called the 'adjunct' mass. The initial acceleration of the bubble is thus $a_0 = F_A/m = g/A$, i.e. it is of the same order of magnitude as g."

"What will happen if we have a small pellet rather than a bubble?" asked Misha.

"In your case, besides the added up mass we have to take into account the mass of the pellet itself, and as well as the Archimedes force we have the force due to gravity — but let us return to our deliberations. As the bubble speed grows, even at vanishingly small viscosity we sooner or later have to consider friction. This resistance force, unrelated to viscosity, should be considered together with the formation of vortices which are being stripped from the bubble. Therefore, the bubble leaves behind it a 'wake' with vortex movements. This also consumes some energy. The speed of liquid in this wake is proportional to v, and its cross section is proportional to the square of the bubble's radius r^2. As the bubble rises through height L, it loses kinetic energy due to the wake of:

$$E_W = Cr^2 Lv^2 \rho$$

Here C is some numerical constant and ρ is the density of the liquid. To generate this energy it is necessary to apply work. Obviously this work is equal to LF_D since the resistance force generates the wake. Hence

$$F_D = Cr^2 v^2 \rho$$

Substituting this value into Archimedes' force, we find the steady state speed of the bubble in a liquid with very low viscosity:

$$v = K\sqrt{gr}$$

Here K is some constant which can be expressed in terms of C. The relationship $v \sim \sqrt{gr}$ could have been obtained immediately from dimensional analysis. Indeed, if the gas viscosity and density in the bubble are vanishingly small, then the steady-state speed of the bubble can only be expressed in terms of g and r — we can produce only one expression with the dimensions of speed: $\sqrt{gr}$. The bubble only accelerates over a distance of several of its diameters before it reaches this constant speed and rises evenly.

"It's not unintentionally that I paid so much time considering the motion of bubbles, because now we will turn to a liquid with anisotropic mass and all that we've learned will be very helpful."

"But please, which of the bubbles moves faster — air or hydrogen?" inquired Misha.

"I do not know for sure myself," answered L.A. "I simply wanted to tell you that the speed difference due to the lightness of hydrogen is negligible. A more significant role is played here by the surface boundary of the gas-liquid system. Different gases do indeed cause changes to the physical properties of the gas-liquid boundary surface, which can affect the bubble movement, but this question will take us far from the matter under consideration.

"Now let us consider a liquid which has a mass density $\rho_\perp$ across the magnetic field that is significantly higher than its normal

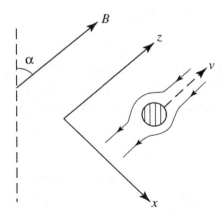

Fig. 6.2   Bubble motion in a liquid with strong mass anisotropy.

density $\rho_\parallel$. We assume that the value $\sqrt{\rho_\perp/\rho_\parallel} = \sqrt{1 + b^2} \simeq b$ is significantly higher than unity. Let the bubble rise in this highly anisotropic liquid with the magnetic field directed at some arbitrary angle to the vertical (see Fig. 6.2). As the liquid is inertially lighter along the magnetic field lines, the streaming around the bubble will proceed in a way such that the current lines will be pressed towards the magnetic field lines. Consequently, it will be easier for the bubble to move along the magnetic field as is shown in Fig. 6.2, but can we be more precise in determining the properties of the stream?

"As it happens, we can. The fact of the matter is that the movement of liquid, although it is more complex than the movement of solid bodies, is still a mechanical movement and so can be quite easily described theoretically. To do so, it is necessary to break this liquid into a multitude of liquid particles and solve Newton's equation for each of them, taking into account the force of gravity and the interactions between the particles. As a result, we obtain the 'Euler' equation. The more complicated generalization of this, which includes the effects of viscosity, is called the Navier-Stokes equation.

"It turns out that the Euler equation for anisotropic mass liquids may be easily transformed into that for isotropic mass liquids. To do so, it is enough to introduce a new coordinate $z' = z/b$ in place of coordinate z along the magnetic field. Moderate changes of z' will then correspond to larger changes of z. In other words, going from z to z' reduces the scale along the magnetic field. The scale for the speed must be changed too: $v'_\| = v_\|/b$. Expressed in terms of the variables z', $v'_\|$ the equation has the usual Euler form (i.e. for isotropic liquid of density $\rho_\perp$), but the expression for the gravity force changes. The new force $\rho_\| g'_\|$ is equal to $\rho_\| b g_\| = \rho_\perp g_\|/b$. Since $b \gg 1$ the acceleration $\boldsymbol{g}' = b\boldsymbol{g}_\| + \boldsymbol{g}_\perp$ in the new variables is close to the magnetic field direction. In the new variables, the bubble also changes its geometry. It starts to resemble a flat cookie oriented in such a way that its axis of symmetry is directed along the magnetic field. Now we can explicitly describe the bubble movement in the anisotropic liquid. For this, let us investigate the bubble movement in coordinates x', z' and to be 'safe and sound' let us rotate the coordinate system in such a way that the effective gravity force is directed downwards (Fig. 6.3).

"One may show that in this case the horizontal plane in the turned coordinates x', z' coincides with the horizontal plane in coordinates x, y. The bubble in coordinates x', z' looks like a flat cookie oriented across the magnetic field. It will not rise straight upwards but will deviate slightly, the deviation being small, since the magnetic field in the coordinates x', z' is directed close to the vertical. Now we can

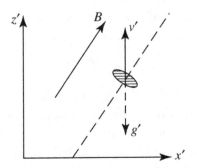

Fig. 6.3   The air bubble in the frame of reference with the artificial coordinates x', z'.

again use Newton's law:

$$ma = F_A - F_D$$

where a is the acceleration, m is the adjunct mass, F_A is Archimedes' force, and F_D is the resistance force. The adjunct mass is proportional to $\rho_\perp$, but not to ρ since the simulated isotropic liquid has density $\rho_\perp$. As this mass is found from the kinetic energy of the liquid streaming around the body, it should be proportional to the volume of the streaming liquid. This volume is certainly larger than the volume of the body because the liquid streams around a thin disk of $r/b \ll r$ thickness, and the streaming volume is $\sim r^3$. Taking account of these considerations, we write the following expression for the adjunct mass:

$$m = A'\rho_\perp V b$$

where the coefficient A' differs somewhat in value from A due to the change in geometry, but still remains of the order of A. The Archimedes force in the auxiliary coordinate system is equal to $F_A = \rho_\| g' V \simeq \rho_\perp g_\| V/b$. It can be seen that in our auxiliary coordinate system the acceleration of the bubble without a resistance force is approximately equal to $\sim g_\|/b^2$.

"The resistance force F_D also contains the density $\rho_\perp$, i.e. $F_D = C' r^2 (v')^2 \rho_\perp$ where $C' \sim C$. Equalizing the Archimedes force and the resistance force and knowing that the disk volume is b times smaller than the volume of the ball, we find the steady speed of the bubble in our auxiliary coordinate system:

$$v' \sim \sqrt{g_\| r}/b$$

Now let us return to the original coordinate system, i.e. to the anisotropic liquid. To do so, Fig. 6.3 should be turned in such a way that the magnetic field returns to its initial position, and then the resultant picture should be extended along the magnetic field b times. The speed and the acceleration of the bubble will increase accordingly (see Fig. 6.4).

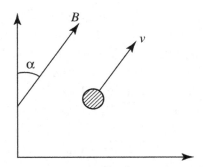

Fig. 6.4   The air bubble in the usual frame of reference.

"Multiplying the above-found auxiliary values a' and v' by b, we find that the bubble in the anisotropic liquid initially has much lower acceleration $a_\| \sim g_\|/b$, but the resultant steady speed is of the former order of magnitude $v \sim \sqrt{g_\| r}$. Note that $a_\perp \sim g_\perp/b^2$ is significantly lower than $a_\|$ if the magnetic field does not 'stick' to the horizontal direction.

"Indeed, our deliberations are very rough, but nevertheless they grasp the main effects of anisotropy. The bubbles don't rise completely vertically, but instead follow an inclined line almost along the magnetic field lines. The increase of the transverse mass density leads to a situation in which the initial acceleration of the bubble significantly decreases, but the final speed remains the same. This happens because the liquid mass density is of great importance for the acceleration value, so that the transverse mass increase slows down the whole process.

"And why does the final speed depend so weakly on the anisotropy? Because the manner of streaming around the bubble changes: the streaming region becomes elongated along the magnetic field so that the longitudinal dimension becomes b times larger than the cross dimension. Therefore the resistance force does not increase significantly.

"Thus, the bubbles in anisotropic liquid rise 'inclined' almost along the magnetic field direction.

"Now, what happens with raindrops? Their acceleration, of course, is not significantly affected by air anisotropy because air is three orders of magnitude lighter than water. If the drops have anisotropic mass they will fall along the magnetic field like the balls discussed previously. The speed of the rain drops reaches a steady value due to air resistance. This speed will be directed almost along the magnetic field if $b \gg 1$. Indeed, at a given speed of a drop the resistance in the longitudinal direction will be significantly lower than that in the perpendicular one. Therefore, the drop will generally move along the field, slowly sliding in the transverse direction.

"And now, Misha," said L.A. "we know quite enough about liquids and gases with anisotropic mass. It's high time to come to reality. I am not going to show you bubbles in the anisotropic liquid — they behave exactly like they should according to our deliberations. Let us refer instead to fountains. Our computer stores quite a number of movies. For example, there is a movie shot in Peterhof that demonstrates its famous fountains. We can store into the computer's memory a program which, instead of ordinary water, will show strongly anisotropic water flows."

L.A. played with the keyboard and suddenly the hall was flooded with the bright and vivid colors of the summer park of Petrodvoretz, and the air was filled with the sounds of falling water from numerous fountains.

"We are now switching on the magnetic field," said L.A. and played with the keyboard again.

The fountains simultaneously twisted and the falling water jets inclined so sharply that Misha grabbed the arms of his chair — it seemed as if the floor was moving away.

"The fact that the falling jets twisted somewhat is quite understandable," said L.A. "All freely falling objects with strongly anisotropic mass move almost along the field. Tiny water droplets fall in the same direction when they move in air with anisotropic mass density. But let us consider the lifting part of the fountain.

"Let us start with the vertical or, to be more precise, formerly vertical fountain. The water with anisotropic mass which rises up inside the tube resembles a string of beads where one follows the other. We are already familiar with the movements of each of these beads. Let us remember that the speed is directed vertically, but the momentum in an arbitrarily inclined magnetic field is directed not vertically but almost across the magnetic field because according to our assumption $\rho_\perp \gg \rho_\parallel$. When anisotropy is switched on, the vertical fountain decreases drastically in height and its jet starts to deviate from the vertical direction due to the fact that the rising liquid has a greater lateral momentum component.

"If prior to applying the field the fountains were slanted, then after the application of the field the following happens to them. The fountain that was initially directed across the magnetic field shortens whereas the one directed along the field becomes longer and higher compared to the situation without the magnetic field, as is shown in Fig. 6.5a,b.

"We can reconstruct the form of the anisotropic fountains more precisely than outlined in these qualitative deliberations. To do so, we have to change the system of coordinates and deal with an isotropic liquid. As we know, the scale along the magnetic field must be reduced b times and the longitudinal component of the gravity force must be increased b times. After this the tubes will turn around and the horizontal plane will be inclined somewhat, but the force

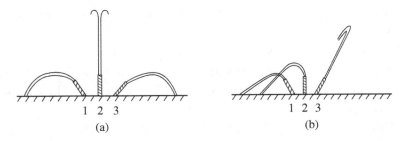

Fig. 6.5 Fountain formation in the case of isotropic mass liquid (a) and liquid with very strong mass anisotropy (b).

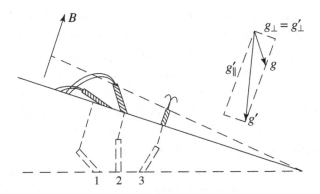

Fig. 6.6   The anisotropic mass fountains reconstructed with the help of the picture plane transformation.

of gravity will again be directed perpendicularly to the transformed horizontal plane (Fig. 6.6)."

"As we see, tubes 1 and 2 are inclined to the new horizon and the inclined fountains will be thrown away from them. Tube 3 is almost vertical in the new system of coordinates, i.e. it is directed orthogonally to the new horizontal plane. If we now stretch the produced picture b times along the magnetic field direction, then the 'horizon' and tubes will come to their former position and the fountains will look as they are shown in Fig. 6.5b.

"A good computer," said L.A. "can easily perform these transformations with jets and show the consequences of anisotropic density effects.

"Now there is another interesting thing in store for you," said L.A.

He pressed some keys on a keyboard and the monitors showed the tropical sea shore that Misha had seen before. The waves were gently licking the sand, and the air was filled with the mild whispers of palm trees.

"Now let us apply a strong magnetic field," said L.A. and in a moment a miracle happened: the waves suddenly changed. They became much higher and moved towards the shore in tight formation.

"What is that?" asked Misha.

"Well, the same story," said L.A. "This is the result of strong mass anisotropy and we can easily understand that the new waves should differ from the ordinary surf. Let us again use transformations in the auxiliary coordinate system or, to put it better, let us reconstruct the behavior of waves in the anisotropic liquid from the behavior of ordinary waves.

"Let the waves coming to the shore look as in Fig. 6.7. These are familiar waves in isotropic water. Now we apply a strong magnetic field and perform a transformation to the auxiliary coordinate system. To this end, we compress the whole picture along the magnetic field direction b times and increase the longitudinal component of gravity b times. The result for a calm sea is shown in (Fig. 6.8). For convenience, we rotate the picture so that the sea is again in a horizontal position (Fig. 6.9). Now in this new 'heavy' sea whose liquid is b times denser than in the normal one, we reconstruct the waves (see dashed line in Fig. 6.9).

"The heavy sea, as we see from Fig. 6.9, is very shallow. Therefore, the waves coming to the shore, which have a height comparable to the depth, are much smaller in absolute amplitude. If we imagine the natural scaling of sea waves, then it is reasonable to assume that the distance between wave crests decreases. If we allow that the transverse scale, i.e. distance between crests, also decreases b times, then the wave period in the 'heavy' sea will be almost similar to

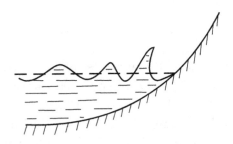

Fig. 6.7   The waves in the isotropic mass sea.

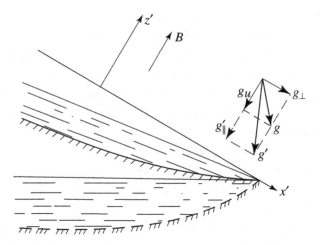

Fig. 6.8   Transformation of the plane to find the shape of waves in a strongly anisotropic liquid.

Fig. 6.9   The wave shape in the strongly anisotropic sea in the artificial frame of reference with x', y' coordinates.

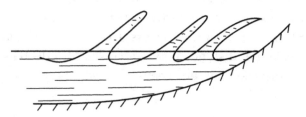

Fig. 6.10   Waves in the strongly anisotropic sea.

that of the ordinary sea, but the waves' phase speed will slow down significantly: the waves will slowly approach the shore.

"Now let us perform the inverse transformation to the ordinary coordinate system by turning Fig. 6.7 to the inclined position as in Fig. 6.8 and then stretching it along the magnetic field b times. What we will see is shown in Fig. 6.10.

"You can see that all the waves are inclined forwards. This happened because that was our choice of the magnetic field direction. Rotating the field anticlockwise, the waves could be set up vertically, or could even be inclined backwards. Nevertheless, the crests of the crushing waves will form from the front. The wave phase speed slows down: highly inclined forward and backwards waves will slowly proceed to the shore and crash in such a way that the splashes falling in front of the waves will slide along the magnetic field direction.

"Well, that is enough for today," said L.A. "Come again in a week's time."

"L.A.," Misha interjected. "I have a good friend, Sasha, who is studying in the physics department of the University. Can I bring him with me? He is a third-year student."

"Then pass him this problem and next time come together."

The professor handed him several sheets of paper with something written on them that looked completely incomprehensible to Misha, but when Sasha saw the pages, he immediately understood what was written on them and started to explain it. (See Problem No. 4 in the Appendix).

7

In the Country of Anisotropic Masses

Next time Misha visited L.A. together with his friend Sasha. L.A. was very polite with the boys and then, kidding, he said:

"Well, well, professors. I would like to take you both on a journey. Please, follow me."

He approached a door in his office which Misha had never noticed before. The boys followed the Professor into a medium-sized room, which was decorated like the hall where L.A. performed his experiments. The room was round and the walls and the ceiling merged into a white dome. There were three chairs in the middle of the room; a steering wheel, like the one on a plane, was located in front of the middle chair. The steering wheel was equipped with a number of strange buttons on an instrumentation panel.

"Misha, can you drive?" asked L.A.

"Just a little," said the boy and flushed. He could not drive and did not want to. He hated noisy city streets filled with animal-like automobiles which seemed to always be trying to hit somebody. It was much safer and more pleasant in the country: no fuss, no terrible noises, no cars.

"That's fine. If you take the steering wheel, you can drive the three of us into our adventure. Sasha and myself will be observing everything attentively. It is much easier to run this machine than a plane or a car.

"We are now under a dome which is made of displays controlled by a supercomputer. The displays are stereoscopic, which is why you will feel as if you are moving on an open platform. Do not be afraid of falling in a hole or hitting a tree — this is just an illusion and poses no danger at all."

Misha sat in the middle chair and took the steering wheel; L.A. sat to his right and Sasha to his left. There was a computer keyboard in front of L.A. and an instrument binnacle in front of Sasha. L.A. played with the keyboard and all of a sudden everything changed. The three of them found themselves in the country outside Moscow: cows were grazing in a field only a kilometer away from them, and beyond that a blue river and a forest could be seen. A small village with a church stood on the steep bank of the river. It was a little cloudy, but still bright. Misha felt that he could smell the fragrance of the flowers and grass.

"Misha, you can now start our journey. The round floor of the room will seem to be an open platform to us, which can move in any direction. By pressing the pedals you can speed up or slow down, and you can turn the steering wheel as if you were driving a car. If you push the steering wheel forwards, we will go down into the ground or into the water; if you pull it back, we will fly into the sky.

"I would like to ask you to remain seated no matter what happens. Just remember that this is all a stereoscopic simulation. To tell you the truth, when I was your age and first found myself watching a stereoscopic movie, I almost hid under the chair when a huge fire engine approached me spraying jets of water. Well, Misha, go ahead and drive and Sasha and myself will watch everything carefully.

"Wait a minute, I have to say something important to you. We will be travelling in a specific country, an anisotropic one. Our super-computer will have to calculate what's going to happen to all natural phenomena if the mass of everything becomes anisotropic. To be more precise, it will have to calculate what will happen if the longitudi-nal and gravitational masses retain their values and the lateral one increases several fold. The computer will take the Earth's magnetic field direction as the anisotropy direction.

"As you know, a compass needle points from south to north, but this direction corresponds only to the component of the magnetic field parallel to the Earth's surface. On Earth, the magnetic field is only parallel to the surface at the equator; it has a perpendicular component at all other latitudes. At the poles, it is the perpendicular component that dominates — the horizontal one is equal to zero. At the latitude of Moscow the magnetic field is directed at an angle of 55° to the horizon. It is along this direction that the former value of the inertial mass is maintained; in the lateral direction the mass of all the material bodies and substances is assumed to be significantly larger than in the longitudinal one.

"Our supercomputer has just completed its calculations, so now let us get acquainted with the results of the computer modeling."

Misha and Sasha listened to the Professor attentively and tried to observe everything around them. At first sight, everything seemed as usual. A falcon was circling somewhere high in the air. Cows were grazing in the field, grasshoppers were chirping in the grass and a plane was droning in the sky.

Suddenly all three of them looked up: they heard an engine sound and saw an airplane approaching them at low altitude. The plane had hardly passed over them when the engine drone started to decrease. From his experience, Misha knew that when it comes to jets you first see the plane itself and then, with a considerable delay, you hear the sound, but here everything was different: it seemed that the sound flew in front of the airplane. Misha and Sasha simultaneously looked at the Professor for an explanation.

L.A. thought a little and then said: "I think that I understand what surprised you. We are in a world where all masses are anisotropic, including air molecules. In the direction across the magnetic field the mass of any molecule significantly exceeds its longitudinal mass. Consequently, the thermal motion of a molecule is much faster in the longitudinal direction than in the transverse one. This means that the speed of sound has different values for longitudinal and lateral directions; their ratio is inversely proportional to the square root of the mass ratio: $c_{s\parallel}/c_{s\perp} = \sqrt{m_\perp/m_\parallel}$. Here c_s is the speed of sound, and the subscripts $\perp$ and $\parallel$ refer to the longitudinal and lateral directions, respectively. Let the plane be at point $A\ldots$" with these words L.A. drew a point on a sheet of paper and then wrote $\boldsymbol{B}$ to show the magnetic field direction (Fig. 7.1).

"If $m_\perp \gg m_\parallel$, the sound will propagate not as spherical waves but as elliptical ones (see Fig. 7.1). It runs faster along the field and slower across the field.

"As a result we have two effects. One of them resembles an ear-trumpet as was once widely used by deaf people. They attached this trumpet to their ear and directed it towards the speaking person. From Fig. 7.1 you can see that the sound energy propagates primarily along the field, like in the ear-trumpet. Therefore, the sound in the lateral direction will be more muffled. The second effect is the non-sphericity of the waves, which leads to the apparent displacement of the sound source. When we try to determine the source of a sound,

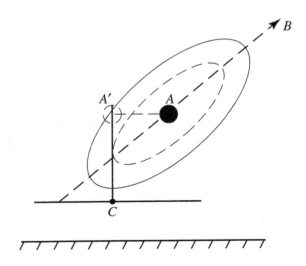

Fig. 7.1   Sound propagation in anisotropic air.

we turn our heads until the sound simultaneously reaches both ears. In other words, we find the direction of the normal to the front of the propagating sound wave.

"Let us assume that we are at point C (Fig. 7.1). If we find the position of the source from the normal to the wavefront, we think it to be at A' instead of A. If the source is a plane, for example, flying in Fig. 7.1 from the right to the left, then it seems to us that the sound source is flying in front of the plane. At Moscow's latitude, the magnetic field is inclined to the horizon, and the source displacement is noticeable when the plane flies from the north to the south. On the contrary, if the plane flies from the south to the north, i.e. from the left to the right in our Fig. 7.1, we will again think point A' to be the sound source, but now this point lies behind the plane, which is at A.

"Thus, it is more difficult to get your bearings in an anisotropic atmosphere. Any time you want to do this, you have to check the direction of the magnetic field. Birds are said to have their own internal compass, but, as we are not birds, we have to follow the compass reading on our instrumentation.

"Well, now let's go somewhere — to the village, perhaps."

Misha pressed the accelerator and slowly steered the platform towards the village.

"L.A., if the air is of anisotropic density, then shouldn't the shape of the plane also be altered in order to account for this effect?" asked Sasha.

"You are quite right, but you won't gain too much. An airplane creates downward-directed air jets. The airplane transfers the vertical momentum to them, thus producing lift.

"In anisotropic air, these jets will generally be directed along the magnetic field, i.e. at some angle to the vertical, but the momentum will again be transferred in the vertical direction. Remember how the ball with anisotropic mass fell under the force of gravity? It fell at an angle, but gained only a vertical momentum component. Thus, in the case of a plane in anisotropic air, the lift is still directed vertically, but is produced by air jets along the magnetic field, i.e. with the same density.

"We can also gain a little by changing the shape of the wing. The vortex flow around a wing in an anisotropic gas differs somewhat from the vortex flow of an isotropic gas. Therefore, the wing shape should be optimized for the new vortex pattern."

So conversing, they approached the village.

"Look!" exclaimed Misha. "The roofs of all the houses are crooked."

"This is easy to explain," responded L.A. "As we know, the raindrops, having anisotropic mass, should fall at some angle along the magnetic field direction even when there is no wind. The wind, of course, can introduce some deviations, but in general the rain direction is given by the B vector. And how would you build a house knowing that the rain will always slant from north to south? You would probably make the northern side of the roof much broader. That's how these houses are built."

The young people noticed that, indeed, all the houses were oriented in such a way that the northern slope of the roof was much broader than the southern one.

L.A. went on, "Moreover, I am sure that the internal layout of the houses is almost the same. As you remember, the water from taps should fall along the magnetic field. It is obviously more convenient to have a tap where the water runs towards you at some angle, not away from you. Thus, taps and showers should be located on the north walls of the rooms."

"Look over there," said Misha. "a gardener is watering his vegetable patch."

They all turned to see a villager watering his vegetables. One end of a long tube was connected to the hose and the other end had a strainer. The man was lifting the tube and then lowering it again.

Now everything was clear without explanation. The drops of water followed the exact direction of the field and fell to the ground along B.

"Look how that little dog is running," said Misha.

The dog was running in a strange sideways manner, placing its paws at an angle to the ground.

"The dog also feels the anisotropic mass of its paws," explained L.A. "See how she slowly moves her paw forward and then quickly places it to the ground not vertically but along the field. It is easier for her to run in such a way as she feels that in a certain direction her paws are less massive. Anyway, let us move farther, say, to the river."

Misha carefully brought the platform to the bank and then slowly placed it directly on the water. The mirrorlike surface of the water surrounded them.

"Strange, the river is very narrow but the current is very slow," Sasha said as if speaking to himself.

"Yes," said L.A. "The current is really very slow. However, the current depends on the river's depth, not its width."

"How come?" asked Sasha.

"Let me explain it to you," answered L.A.

"Let us denote the river's depth by h and its width by l. Usually the depth is much smaller than the width, i.e. $h \ll l$, so the effect of the banks on the current is practically negligible. Now let us try to find the dependence of the speed of the water v on the river depth and on the angle of the slope γ (see Fig. 7.2). It is easy to see that the speed is determined by this angle. The driving force of the flow is the component of acceleration along the river surface. This component is $g \cos \gamma \approx \gamma g$ when γ is small.

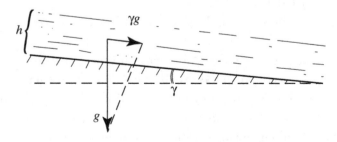

Fig. 7.2   Water flow under the influence of the longitudinal component $\gamma \rho g$ of the gravitational force.

"Now, how do we find the dependence of the flow speed on γg? Here we can apply the general but effective method of 'dimensional analysis'. γg has the dimensions of $\sec^{-2}$. We want to produce a value with the dimensions of speed, $\mathrm{m\,sec^{-1}}$, so we should proceed as follows. We take the square root of γg and multiply it by a value with the dimensions of length. Following this method, we express the speed as:

$$v = Ch\sqrt{\gamma g}$$

where C is a dimensionless constant and h is the river depth. As we see, the value v has the dimensions of speed. With the appropriate choice of C, we obtain the real speed.

"Thus, the flow speed on a stretch of river with a given slope is proportional to its depth. As the slope steepens, the speed increases with the square root of the slope angle. So, why do we often relate the water speed to the river's width? The fact is that such an impression is formed when we observe the same river on both wide and narrow stretches. The total volume of water flowing past a given point per unit time is obviously equal to $\mathcal{P} = vhl$, where l is the river's width. On a stretch where there are no tributaries feeding into the river, $\mathcal{P}$ can be considered constant. Therefore, in places where the river is broader, its depth and speed should be lower. Although the river speed is determined exclusively by its depth and slope, the indirect dependence of its depth on its width creates the illusion that the speed of the river depends on its width.

"Until now, we've been thinking of an ordinary river with isotropic mass. If the mass is anisotropic, and the lateral mass is many times higher than the gravitational mass, the river speed will become much slower. It will be proportional to the square root of the ratio of the longitudinal mass to the effective mass as we can see from our discussions while observing the fall of the bead. Clearly, in the country of anisotropic masses the river speed is significantly lower than in the isotropic case."

"Look what a steep bank this is," exclaimed Misha.

"Now we will check the compass," responded L.A. "That is right — this is the northern slope. That is why it is so steep. Usually, the slopes of steep banks, except at the very upper wall, look like screes. Their slope is determined by the angle at which small pieces of earth and sand slide down. Here, however, we have a northern slope in the country of anisotropic masses. As we know, here objects don't fall vertically, but at a certain angle almost along the magnetic field direction. At the northern slope the magnetic lines are inclined such that the lower part of the line is displaced towards the river. Soil slides are therefore impossible — small stones and pieces of earth immediately fall outwards and down, and this is the explanation of the steep slope."

"I have noticed that the clouds here move much more slowly than normal," said Sasha.

"Yes, certainly. The winds here are very slow due to the increase of the lateral mass of the air molecules. However, I have not yet quite understood the details of the effect."

"But why does the wind blow?" asked Misha.

"This is not an easy question, but let us try to understand the phenomenon," answered L.A. "Ultimately, winds are developed as a result of air convection. The surface of the Earth is heated by the Sun, and a shallow layer of the atmosphere — about a hundred meters thick — becomes superheated. This layer is much lighter than the layers above and so convectional air lift occurs from time to time. This phenomenon is easy to observe in the early hours of the day in a field where the process of cloud formation is beginning. If you are attentive, you may notice that the cloud forms unexpectedly, as if out of nothing. A parcel of air is heated and then rises through air layers of steadily decreasing pressure. The parcel expands, which causes its temperature to decrease. At a certain height the air temperature reaches a point where the water vapor starts to condense, forming tiny droplets of fog. This is the height at which the cloud starts to develop. All cumuli have a flat base as if they were cut from the bottom: they are sitting at this condensation-forming altitude.

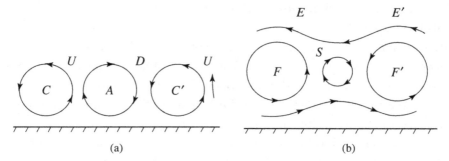

Fig. 7.3 Convective cells in the atmosphere (a) and their merging (b) which leads to the formation of winds.

"Now let us examine these air flows in vertical cross-section. In Fig. 7.3a we observe two rising air flows at points U, U' and between them a descending flow D, which is pushed down by the rising flows. At points U, U' clouds may be developed, but for now we will concentrate on the structure of the flows rather than on clouds. As we see, they resemble a chain of vortices with alternating signs: at points C and C' the air rotates anti-clockwise, and at point A it rotates clockwise. In the very upper layer we have an air flow with alternating sign; the direction of the air flow changes depending on the point of observation.

"It turns out that this periodic structure is unstable: it tends to grow into a unidirectional flow.

"Let the central vortex be less intense than the adjacent vortices (Fig. 7.3b). A part of the flow in the upper part of the vortex, i.e. along the line EE', will shift from one outer vortex to another outer vortex. As a result, at point S the oncoming flows will be close to each other. Due to the action of viscosity, the inner vortex will slow down and its intensity will decrease. If the central vortex vanishes, the flows will meet at points F, F', and will break each other. The resultant flow will be divided into two flows. The ground flow, directed in Fig. 7.3b to the right, tends to decay due to viscous friction with the Earth's surface. The upper layer EE', which is moving to the left, generates wind.

"Thus, convection inevitably creates winds at an altitude above about 1–2 km. If observed from space, these winds resemble small vortices when they form. The vortices, however, tend to merge, forming large-scale vortices easily seen from space. These gigantic vortices develop into cyclones and anticylones. The air in cyclones rotates in the direction of the Earth's rotation, whereas the air in anticyclones rotates in the opposite direction.

"The main wind which determines the flow of masses of air is called the geostrophic wind. It is interesting that the wind blows not from high-pressure regions towards lower pressure zones, but almost perpendicular to the pressure drop. (The pressure drop is a vector proportional to the rate of pressure increase per unit length in the direction of pressure increase).

"This happens because the pressure drop is balanced by the 'Coriolis force'. The Coriolis force is proportional to speed, and is directed perpendicular to the velocity: to the right in the northern hemisphere, and to the left in the southern hemisphere. It is the Coriolis force that equalizes the pressure drop so that the wind blows almost along isobars — lines of constant pressure."

"Now we can consider what changes in these processes will be introduced by mass anisotropy. First of all, the convection pattern will change: upwards and downwards flows will be directed, as a rule, along B, i.e. at some angle to the Earth's surface. Besides that, the transverse convection scale will decrease due to the fact that the transverse mass will have increased. As a result, the transverse speed in the convection cell will noticeably decrease, so the wind generated by these cells will weaken. Consequently, the strength of the geostrophic wind, which is responsible for large-scale weather changes, will decrease. The Coriolis force will be affected by two changes with different signs: because of the speed decrease it will decrease, and because of the growth of the transverse mass it will increase. The pressure distribution in the atmosphere will adjust itself to the new wind patterns. The main thing that will happen is a significant decrease in wind speed. However, the destructive force of hurricanes and tornadoes will remain unchanged because the transverse mass will grow."

"L.A., can I ask you a question?" asked Sasha.

"Of course," answered L.A.

"Recently, when I was flying to Vladivostok, the air stewardess announced that the outside temperature was minus 45°C. How can that be? The cold air should sink and be replaced by warmer air."

"Your question is correct, but I have omitted a very important thing. When a parcel of warm air rises, it expands because the atmospheric pressure decreases with altitude. While expanding, the air cools down, since there is not enough time for heat to be transferred between molecules before the expansion drives them apart. This process of expansion without heat exchange is called adiabatic expansion. Therefore, if you imagine a rising parcel of air, its temperature will be different at different altitudes: the higher the altitude, the lower the temperature. This sort of temperature distribution is called 'isentropic'.

"An atmosphere with such a temperature distribution doesn't convect. Convection starts when the temperature grows a little bit faster with decreasing height. Even with convection the atmospheric temperature decreases with increasing height. This explains why high mountains are covered with eternal snow and glaciers. Well, now let us go closer to the forest."

Misha confidently took the platform from the river and drove it towards the forest, stopping at its edge. At first glance, nothing seemed out of the ordinary.

Sasha was the first to notice something unusual. The leaves were responding to the wind very slowly and soundlessly.

"The complete quietness is quite understandable," said L.A. "We have approached the forest from its northern side. In the anisotropic air the sound is generally directed upwards and is scattered. The leaves, meanwhile, move slowly because of the slower wind and because their transverse mass is higher than the regular isotropic one. Therefore, if the graft elasticity is intact, the oscillation frequency of

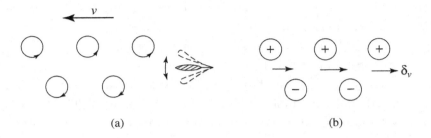

Fig. 7.4   Leaf flutter (a) and the series of vortices produced (b).

a leaf will decrease. The trembling of a leaf in the wind is called flutter, and now I will explain its origin to you. Let the leaf be trembling
as shown in Fig. 7.4a with the wind blowing to the left.

"At each flap the leaf excites a small vortex, and a double vortex
chain moves leftwards from the leaf — two small vortices after each
complete flap of the leaf. The upper and lower chains of vortices in
Fig. 7.4a have alternating signs. Therefore, as is shown schematically
in Fig. 7.4b, a small counterflow with velocity δv develops between
the two vortex chains.

"As the wind speed is somewhat lower in the region between two
alternating sign vortex chains, the total kinetic energy of the wind
with vortices is also lower than it was prior to their development. One
can say that in a fixed coordinate system at rest a double vortex chain
possesses negative energy. It is necessary to emphasize here that we
are speaking of the kinetic energy in the frame of reference at rest.
If we use the coordinate system that is stationary with respect to
the wind, then we will see that before the generation of vortices the
kinetic energy was equal to zero, and after their development the
energy increased.

"Let $\mathcal{E}$ be the positive energy of the vortex chain per unit length
calculated in the coordinate system of the initial wind. In the same
coordinate system, the vortex chain has momentum P per unit
length. As seen from Fig. 7.4b, this momentum is related to δv and
is directed to the right. The energy $\mathcal{E}_0$ of the vortex chain in the

coordinate system at rest is equal to:

$$\mathcal{E}_0 = \mathcal{E} - Pv$$

The second term on the right has a negative sign because the wind speed and the vortex chain momentum have different directions. If the wind speed is high enough the energy $\mathcal{E}_0$ becomes negative. In this case, in order to excite the vortex chain you do not have to introduce energy, but take it away from the wind, so to speak. This is just how flutter begins: the leaf exciting the vortex chain receives energy from the wind, which is spent to sustain its oscillations.

"This flutter phenomenon is widely observed in nature. One has to take it into account in technology as well: no new airplane could be designed without a thorough investigation of flutter development.

"Well, we have been travelling for quite a long time and I still have plenty to do today."

With these words, L.A. switched off all the monitors, and instead of a new exciting country they again found themselves in a small laboratory with a dome.

"I hope that you now understand mass anisotropy much better," said L.A.

"We certainly do," answered Sasha, "but I cannot understand how the magnetic field can cause mass anisotropy. I haven't come across this phenomenon in any of the books I've read."

"Come to me in a week's time and I will explain the properties of matter in a superhigh magnetic field, and the origin of mass anisotropy to you."

Misha and Sasha said goodbye and went home to talk over the events of the day.

8

Matter in a Superstrong
Magnetic Field

"Today we will get acquainted with the behavior of matter in a super-strong magnetic field" L.A. began, when Misha and Sasha next came to him.

L.A. approached a large whiteboard and started his narration.

"We shall begin, naturally, with the simplest objects. Let us begin with a single electron. Classically, an electron is just a charged particle with charge e and mass m. The charge is equal to $1.6 \cdot 10^{-19}$ Coulombs, and its mass is $0.9 \cdot 10^{-27}$ g. A free electron in classical mechanics moves with a constant speed along a straight line, or rests if its velocity is precisely equal to zero."

L.A. took a blue pen and drew a point and an arrow on the board showing the direction of the electron.

"L.A.," interjected Sasha. "What is the color of an electron? You have drawn it blue and somewhere I saw it drawn orange. What is the real color of the electron?"

"I will gladly answer your question," said L.A. "Don't hesitate to ask me questions and interrupt me at the most unsuitable moments. My purpose in talking with you is to hear your questions. So, what is color? Let us start with black and white. We see white where there is a surface that scatters all the incident light without any absorption. Conversely, black surfaces absorb all incident light.

"In between these extremes, there are a large number of shades of gray, when the surface partially absorbs the incoming light, and the rest is scattered evenly over the spectrum, i.e. without emitting any color. Color develops when the reflected spectrum differs from the incoming one. As you know, natural sunlight, or the white light scattered from clouds, actually consists of many colors, which we conventionally call red, orange, yellow, green, blue, indigo and violet when we see them spread out in a rainbow.

"In reality, the dispersion into seven colors of the rainbow is very arbitrary, but more or less acceptable to all of us. If a surface scatters light unevenly over the spectrum, then it appears to be colored. For instance, if a rose petal absorbs all the colors of the rainbow except red, then we see a red flower. Or, if air generally scatters the short-wave part of the spectrum, then we see a blue sky.

"Now let's return to the electron. It so happens that the electron evenly scatters all the colors of the visible part of the spectrum. In this respect, it is like the small droplets of fog which form clouds. Therefore, we can be positive that the electron is of white color: a cloud of many electrons would look like an ordinary cloud in the sky. Thus, the electron has a white color, but I cannot draw it white on a whiteboard, so we have to make do with blue.

"As we have started to discuss color, I would like to say a few words about the color of strongly heated objects. I will shortly introduce you to the physics of pulsars, and you will learn that the surface of a pulsar has a temperature thousands of times higher than the Sun's temperature. The question is: what color will a pulsar be?

"It should be noted that the human eye does not see the whole spectrum: for example, we cannot see in the infrared or ultraviolet. Hot objects radiate heat away over a range of frequencies, and the range shifts towards longer wavelengths as the objects cool. For example, iron can be heated until it glows red, but at room temperature it emits mainly infrared radiation. We cannot see this, but some snakes can — moreover, they can see a warm-blooded body against a lower temperature background.

"Conversely, if we start raising the temperature of a body, then eventually the ultraviolet part of the spectrum starts to prevail, which we do not see either. Above a certain temperature, further temperature rises do not affect the color: all changes in the spectrum occur in the invisible region. If we keep the total power being radiated constant (e.g., by means of a gray absorber), then in the visible spectrum the intensity will decrease as the temperature increases. For example, in tokamaks — devices for the production of plasma with temperatures of hundreds of millions of degrees — the central part of the plasma seems to be dark even though it is heated up to much higher temperatures than its iridescent shell.

"Thus, the color of the pulsar is the same as the color of hot stars, i.e. blue. That was why the floor was blue in the hall where we experimented with balls, as it would have been on the pulsar.

"Let us now return to matter in a super-strong magnetic field. As I have said earlier, the classical electron represents a charged point mass. According to Newton's laws, such a point moves with a constant speed unless it is affected by some external force. Let this force be due to the magnetic field.

"It is known that in a magnetic field a particle with charge e is affected by the Lorentz force:

$$F_L = ev_\perp B$$

Here $v_\perp$ is the velocity perpendicular to the magnetic field, and B is the magnetic field strength. The Lorentz force acts in a direction perpendicular to both the particle velocity and the magnetic field vector. It therefore resembles a centripetal force, and the particle moves along a circular orbit of radius $\rho_\perp$. This radius is called the Larmor radius. The centripetal force is equal to $m\omega v_\perp$, where ω is the angular frequency. If we set this equal to the Lorentz force we obtain:

$$\omega = eB/m$$

This frequency is usually called the cyclotron frequency. The Larmor radius is $\rho_\perp = v_\perp/\omega = mv_\perp/eB$. We see that, for a given velocity $v_\perp$,

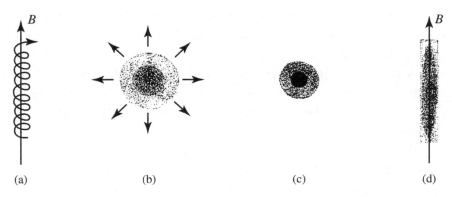

Fig. 8.1 (a) A charged classical particle moves in a magnetic field B along a screw-like trajectory, with the screw-step depending on the particle's velocity along the magnetic lines, and the orbiting radius depending on the transverse velocity and field strength; (b) a free electron looks like an expanding negatively charged cloud; (c) in a hydrogen atom, an electron is attracted by the proton and forms a compact cloud of 10^{-8} cm in diameter (this diameter is identified with the hydrogen atom's dimensions); (d) in a magnetic field, an electron behaves like a cloud that can expand only in the direction of the field, and is confined in a tube with respect to transverse directions.

the Larmor radius diminishes as the magnetic field increases. Meanwhile, along the field direction the particle moves freely (Fig. 8.1a).

"This is what happens with a classical particle, but an electron is not just a charged point mass — it possesses some wave properties. These properties do not allow the existence of a motionless or steadily moving electron of infinitely small dimensions. Roughly speaking, due to its wave properties, the electron represents a charged cloud with a tendency to expand (Fig. 8.1b).

"In an atom an electron is pulled towards the nucleus, and its tendency to expand is compensated for by the electric force (Fig. 8.1c). The natural scale of atomic dimensions is the radius of a hydrogen atom, which is usually called the Bohr radius, $a_0 = 0.5 \cdot 10^{-8}$ cm. As the elements become heavier, the attractive force between an inner electron and the nucleus grows, so the radius of the innermost electron clouds decreases. Additional electrons occupy the outermost shells, and so an atom's size does not change significantly.

"In the presence of a very strong magnetic field the structure of atoms and molecules can change drastically. A classical charged particle in a magnetic field moves along a helical line — its motion across the magnetic field is constrained (Fig. 8.1a). Similarly, the electron cloud in a strong magnetic field may freely expand only along the field; in the transverse direction it is localized in a tube (Fig. 8.1d).

"What values of the magnetic field can be considered to be very strong? By very strong, we mean that the force due to the magnetic field is comparable with the electric forces between the electrons and nucleus. In this case the Larmor radius becomes less than the Bohr radius. Simple estimates show that this happens when the field is $B_0 = 2.35 \cdot 10^9$ Gauss. We will call fields of strength $B > B_0$ superstrong magnetic fields.

"Let us place atoms in a superstrong magnetic field. The simplest atom is the hydrogen atom, which consists of one proton and one electron. The proton, being the heavier particle, represents a charged point while the electron is a cloud stretched along the magnetic field direction (Fig. 8.2a). The electric force does not permit the cloud to expand too far along the magnetic field, but the electric force

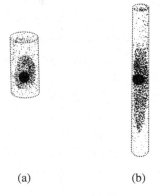

(a)   (b)

Fig. 8.2   A hydrogen atom in a superstrong magnetic field at $B \gg B_0$: (a) at $B = 10^{12}$ Gauss the cross-field/along-field size ratio, δ, of the electron 'cloud' is 3; (b) at $B = 10^{13}$ Gauss, $\delta = 8$. The central bullet is the nucleus.

is smaller than that due to the magnetic field. Consequently, the electron cloud is much more weakly constrained along the magnetic field direction than across it.

"Thus in a superstrong magnetic field the electrons tend to move along the field lines. The stronger the field, the smaller the radius of the tube is (see Fig. 8.2b). For example, at $B = 10^{12}$ Gauss the hydrogen atom in its ground state should resemble a sausage elongated along the magnetic field; the ratio of its length to its width, δ, is $\delta \sim 3$ (see Fig. 8.2a). As B grows, the atom's dimensions decrease. For example, at $B = 10^{13}$ Gauss $\delta \sim 8$: the atom resembles a thin needle (Fig. 8.2b).

"As the magnetic field grows, the electron and proton move closer to each other. Therefore, the binding energy of the atom grows (Fig. 8.3). In normal conditions the energy of the hydrogen atom's ground state is $\mathcal{E}_0 = 13.6$ eV. At $B = 10^{12}$ Gauss, $\mathcal{E}$ is almost 10 times higher than $\mathcal{E}_0$, and at $B = 10^{13}$ Gauss it is almost 20 times higher than $\mathcal{E}_0$.

"The excited levels of a hydrogen atom look completely different in a superstrong magnetic field. Due to the extremely high anisotropy, the excited states of the electron's longitudinal and transverse motions are different. It is much easier to excite the levels

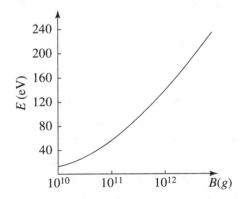

Fig. 8.3   A hydrogen atom's binding energy $\mathcal{E}$ in a magnetic field. To a first approximation it is given by a formula of the form $\mathcal{E} = \mathcal{E}_0(\log B/B_0)^2$ with $\mathcal{E}_0 = 13.6$ eV being the ground state energy in normal conditions.

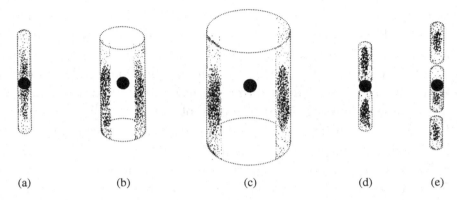

(a)   (b)   (c)   (d)   (e)

Fig. 8.4   To obtain excited states of a hydrogen atom in a superstrong magnetic field,
one has either to increase the 'electron cloud' dimensions or to redistribute the electron
density along the magnetic field. At $B \gg B_0$ (a) the electron cloud in the ground state;
(b) in the first excited state; (c) in the second excited state; (d) in the first longitudinal
state; (e) in an excited longitudinal state.

that are related to motion across the field. The main energy level
(Fig. 8.4a) corresponds to the atom's lowest state. In this ground
state the electron cloud, elongated along the magnetic field, rotates
about the nucleus with a 'drift' velocity. The cloud fills the tube as
is seen in Fig. 8.4b. The tube's radius grows with the growth of the
excitation energy (Fig. 8.4c). The excitation energy of these levels is
comparatively low since the main features of the electron cloud configuration are conserved — it is elongated along the magnetic field,
and the charge density increases whilst approaching the nucleus in
the longitudinal direction.

"In addition, excited states exist that correspond to changes of
the electron density distribution in the direction of the field. The
first longitudinal excited state corresponds to that when the electron
density is equal to zero at the center of the nucleus (Fig. 8.4d). It
happens to have an energy close to the energy $\mathcal{E}_0$ of the hydrogen
atom's ground state in the absence of the magnetic field. In order to
excite this level in the superstrong magnetic field a very high energy
is needed. Subsequent longitudinal levels require smaller amounts of
energy for their excitation.

"Along with these energy levels for cross field and longitudinal motion there are levels with a very high excitation energy. They are related to the quantum motion of a free electron in the magnetic field. They are called Landau levels and correspond to large values of the Larmor radius $\rho_\perp$. However, we are not currently interested in them.

"Now we can proceed with the structure of heavier atoms. The shell structure of heavy atoms in superstrong fields changes significantly. It is convenient to construct these atoms by starting with a bare nucleus and adding one electron at a time in such a way that the energy remains minimal, i.e. so that the ion or atom remains in its ground state. Thus, let us take a nucleus with atomic number Z and place the first electron in the lowest level. This level corresponds to the ground state (see Fig. 8.2). We obtain a hydrogen-like system, with the difference that, due to the larger nuclear charge, the radius turns out to be Z times smaller than the Bohr radius. Let us now add another electron. In normal conditions, two electrons with opposite spins (magnetic moments) can be placed in each energy level. In a superstrong magnetic field, however, all the magnetic moments are aligned along the field, so only a single electron can be placed in each shell. Since the lowest levels are those corresponding to transverse and drift motion, the second electron should be placed on the level drawn in Fig. 8.4b, the next as in Fig. 8.4c, and so on.

"With such a structure the atom expands in the radial direction (Fig. 8.5) and so we have to check whether we can proceed in the same way with all Z levels until we reach a neutral atom. By the way, note that the atom should maintain an elongated shape along the magnetic field. It turns out that we can view the atom as a stick oriented along the magnetic field when the condition $B \gg Z^3 B_0$ is satisfied. The higher the nuclear charge is, the thicker the stick should be.

"Now let us address the case of a very heavy atom when $Z^3 B_0 \gg B$. In this case, whilst filling the levels for transverse motion, we can reach a radius comparable to the height of the cylinder. Further filling of these levels becomes useless, since the corresponding shells are located too far from the nucleus. It is much more profitable to

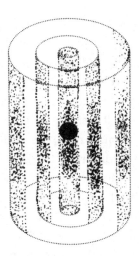

Fig. 8.5   Electron shells for a three-electron atom composed of the coaxial electron clouds presented in Fig. 8.4 a,b and c. For $B \gg Z^3 B_0$ the subsequent electrons will form clouds of the form Fig. 8.4d.

place the next electron on the first excitation level of the longitudinal motion. For heavier atoms it turns out that the central part of the atom maintains almost spherical symmetry. The influence of the superstrong magnetic field in this case is that the radius of the atom is reduced, its external shells become somewhat elongated along the field, and its binding energy increases. Even almost spherically symmetric heavy atoms are still noticeably elongated along the magnetic field.

"Let us assume that all atoms in this superstrong magnetic field are significantly elongated along the field, but each to a different degree. Remember that with increasing Z the radius increases and the longitudinal dimension reduces. In other words, very light atoms resemble long thin needles and heavier atoms resemble short thick sticks (see Fig. 8.6a).

"It is easy to see that such atoms should have very strong interactions with each other — much stronger than those of atoms under normal conditions. Indeed, in normal spherically symmetric atoms the positively charged nucleus is 'hidden' inside the electron cloud,

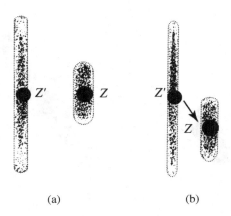

Fig. 8.6   Molecule formation in a superstrong magnetic field: (a) different atoms in a field $B \gg Z^3 B_0$; (b) a molecule formed by a light and a heavy atom — Z' and Z.

and interaction forces arise only due to a weak polarization — deformation — of the cloud. In the atoms shown in Fig. 8.6a,b the nuclei are practically 'naked' — the electron clouds do not cover them on the sides. Therefore, if we take a very light atom Z' and place it close to a heavy one, then the nucleus Z', being almost naked (i.e. with a remote electron cloud), will be very energetically pulled into the dense electron cloud of the Z nucleus. As a result, a molecule is formed: the light nucleus Z' will be held to the heavier nucleus at a distance such that the mutual repulsion of the nuclei balances the force pulling the lighter nucleus into the electron cloud. The electrons of the Z' nucleus will position themselves on the external levels of the molecule (Fig. 8.6b).

"The heavy nuclei may be able to absorb several, let's say n, light atoms to form a molecule. The capture of light nuclei will cease when $nZ' \approx Z$, as here the repulsion of the nuclei will prevail over their attraction by the electron cloud. The total binding energy of such a molecule is comparable to the binding energy of an atom with atomic number $Z + nZ'$. This is certainly a very high value.

"Now let us consider identical atoms. In this case the nearly naked nuclei will strongly interact with the electron 'sticks' of neighboring

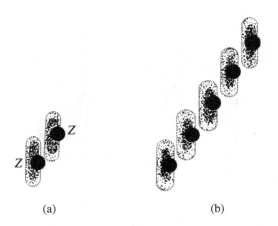

Fig. 8.7 In a superstrong magnetic field a polymer chain appears to be energetically favored over a bulk crystal structure. (a) Two identical atoms glued together; (b) a polymer chain formation.

atoms (Fig. 8.7a). The two atoms will approach each other such that the nucleus of one atom enters a dense part of the electron cloud of the other. However, due to the electrostatic repulsion between the nuclei, they will remain separated along the magnetic field.

"When two similar atoms approach each other, the electron shells never totally combine — only the electrons on the external orbits become common. One may say that atoms maintain their individuality (calculations show that a complete unification of the electrons may also occur, but only in very strong fields).

"If another atom is added to a molecule consisting of two atoms (see Fig. 8.7a), then it will attach itself either to the upper or to the lower atom. This ensures that the added nucleus passes close to the molecule's electron shell, but remains at a large distance from other nuclei. The same will be true of every subsequently added atom. As a result, we obtain a long polymer chain (see Fig. 8.7b). This chain will be somewhat inclined with respect to the magnetic field, and the stronger the magnetic field, the smaller the angle of inclination.

"Figure 8.7 gives only a schematic picture of a polymer. For simplicity we have drawn it with all the atoms in the same plane.

However, this is not the ideal configuration from the point of view of energy. We arrive at the actual configuration by moving every second atom out of the page towards us, and then compressing the molecule across the magnetic field in such a way that the distances between adjacent atoms remain unchanged. Atoms located next to one another are also attracted to each other, so from the point of view of energy it is better not to have a simple linear chain, as shown in Fig. 8.7, but a thicker polymer chain. The binding energy of such a chain is still mostly built up from interactions with the closest neighbors, so it does not differ greatly from that of the simple linear chain shown in Fig. 8.7.

"Let us now consider the behavior of a solid body in a superstrong field. The most typical structure of a solid body in normal conditions is a crystal structure. One could suppose that in a superstrong magnetic field atoms should also form a crystal, but this is not true: the crystal structure is energetically inefficient. This is obvious from Fig. 8.7b, where we added one atom to another and observed that it was favorable for subsequent atoms to attach to the ends of the long chain. In other words, polymer chains are first formed, and later, due to much weaker interaction forces, form a solid body. Thus, in a superstrong magnetic field a solid body should be polymeric.

"Yet this is a rather unusual polymer. The molecules are arranged at small angles to the magnetic field lines, and are elongated along the field. This should result in a number of unusual properties. For example, it is easy to see that it is very difficult, or even impossible, to rotate such a body round the axis perpendicular to the magnetic field; this would require all the molecules to be reconstructed to again lie in the magnetic field direction. We can assume that if a torque were applied to such a body, a breakdown would occur in the body rather than a rotation through the field (see Fig. 8.8a). On the other hand, it is easy to rotate the body about the magnetic field axis (Fig. 8.8b,c).

"If the magnetic field is very strong, much higher than 10^{12} Gauss, a macroscopic body reveals another strange mechanical property which is related to the peculiar features of its atoms. This is

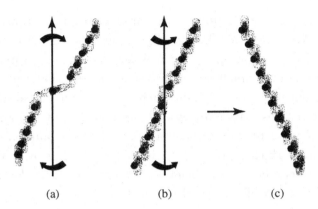

Fig. 8.8 (a) A polymer chain breaking under the action of a torque directed perpendicularly to the magnetic field B; (b) a polymer chain may be easily rotated along the magnetic field axis without deformation.

anisotropic mass. Let us take a single atom and apply a force to accelerate it. If the acceleration is along the magnetic field, nothing unusual happens, but if the acceleration is across the field, the atom's configuration will change. For an atom moving across a magnetic field with velocity $v_\perp$, its motion leads to the existence of a transverse electric field of magnitude $v_\perp B$ in its own frame of reference. This field deforms the electron shells — the electron 'spindle' shifts away from the nuclei and across the field. The binding energy of the atom increases by an amount proportional to the square of the velocity, and this energy increase can be regarded as an increase in the atom's transverse mass. Hence, the mass across the magnetic field of a macroscopic body should be larger than the longitudinal one: it is easier to accelerate a body along the field than across it. Consequently, in a more complex magnetic field with curved field lines, the body 'prefers' sliding along the field lines to moving across them.

"Other physical properties of matter in a superstrong magnetic field should also be different. For instance, we are used to the fact that as temperature increases a solid body starts to melt, then the liquid evaporates and then, at higher temperatures still, molecules dissociate and atoms are ionized, i.e. a plasma is formed.

"In a superstrong magnetic field all this is different. For matter with heavy atoms the binding energy of atoms in a molecule is much higher (approximately Z times) than the ionization energy. It is even higher than the excitation energy of low-lying energy levels. When a body is heated, first the electrons are excited to higher levels and the atoms and molecules swell, then the interatomic bonds are destroyed in molecules. Correspondingly, the conditions of molecules' and atoms' existence can be different.

"At the surface of the pulsar (where the pressure is low and free atoms and molecules could exist) the temperature is about 1 million degrees, i.e. $\sim 100\,\mathrm{eV}$. However, in a field of 10^{13} Gauss, even at such high temperature molecules formed from atoms with moderately large Z could exist. More complex objects of the polymer-type are also possible.

"Thus, a superstrong magnetic field should strongly affect the structure of atoms and molecules. This should also influence the properties of a pulsar's surface, or, to be more precise, its crust. Due to the higher values of the binding energies of atoms and molecules and the binding forces in a solid body, the surface layer will not necessarily be a hot plasma, as it is for regular stars. It could be a hot neutral gas or even solid.

"Thus, extraordinary changes in the properties of matter take place in a superstrong magnetic field," L.A. concluded.

"Now I would like Sasha to attempt another two problems at home and tell Misha about them."

L.A. said good-bye and invited the boys to come and see him in a week.

9

Neutron Star

"Today I will tell you about neutron stars," said L.A. at their next meeting.

"All stars, including the Sun, are huge balls of gas. Or, rather, they are huge balls of plasma since at high temperature the atoms of ordinary gas break down into their constituent nuclei and electrons. In a plasma, the nuclei look like positively charged ions located in an electron gas. However, from the point of view of expanding when heated, a plasma acts like a regular gas. Thus, a star is a ball of gas prevented from expanding by gravity.

"The hot state of this ball is maintained by fusion reactions, i.e. the fusion of light nuclei into heavier ones, which is accompanied by the release of energy.

"The main fusion reaction in the Sun involves the 'burning' of hydrogen into helium. In other words, the Sun is a gigantic fusion reactor with a gravitationally confined plasma.

"A star maintains its temperature and density automatically at the level necessary for the reaction. If the plasma density, for instance, increases, the rate of the reaction increases too, and the star heats up and expands. As a result, the density decreases again. There are situations where a star contracts and broadens periodically along the radius. Very low amplitude radial oscillations can be observed even in the Sun.

"Stars which are 'older' than the Sun have had enough time to burn all their hydrogen; in these stars fusion reactions between heavier nuclei take place.

"Can you imagine the extraordinary situation where all the atomic nuclei of a star are fused into one gigantic nucleus? This is quite possible, and such a star is called a neutron star. Why just neutron, and not neutron-proton, like regular nuclei? I'll explain it to you.

"The fact is that regular nuclei become enriched with neutrons as the atomic weight grows. For example, the nucleus of ^{238}U has 146 neutrons and only 92 protons. This happens because, from the point of view of energy, it is favorable for 'excess' protons to fuse with the electrons of the atomic shell to become neutrons. The same is true for supercompressed matter. In a large volume of neutron star, the number of protons (and electrons) is very small. Neutron stars are held together by gravitational forces.

"One of the first theoretical models of a neutron star was suggested by Professor L.D. Landau. We will not go into the details, but some things should be discussed.

"For example, one can make rough estimates of the physical parameters of the neutron star. A neutron star is an analog of an atomic nucleus while a regular star is an analog of an atom. An atomic nucleus has a radius of order 10^{-13} cm whereas the radius of a typical atom is of order 10^{-8} cm. Thus, their sizes differ by a factor of 10^5. Correspondingly, the radius of a neutron star is approximately 10^5 times smaller than that of a regular star. If the dimensions of the Sun are of order 10^6 km, those of a neutron star are ~ 10 km.

"Similarly, we can try to estimate the period of a neutron star's rotation by comparison with the Sun. The period of the Sun's rotation is about $2 \cdot 10^6$ sec. If the Sun collapsed until its diameter became 10^5 times smaller than at present, its rotation period would become 10^{10} times shorter due to the conservation of angular momentum.

"Thus, in this case one can expect a period of order 10^{-4} seconds. In fact, the periods of rotation of neutron stars are longer: from a

few milliseconds up to several seconds. This means that our estimate
was very rough.

"On the whole a neutron star consists of neutrons. However, it also
contains a small number of protons — analogous to a dilute solution
in a neutron solvent. The relative density of protons is very low: it is
about 10^{-9} the density of neutrons. The relative density of electrons
is equal to the density of protons to a high accuracy. Thus, in a
small volume of neutron star, the number of protons and electrons
is the same. The positive charges of the protons are balanced by the
negative charges of the electrons, so that the star is quasi-neutral.

"Quasi-neutrality does not prevent the existence of currents in a
neutron star. These currents should be 'closed' inside the star. A simple example of a closed-circuit current is a regular solenoid connected
to a current source. If the circuit has a high conductivity then current
can flow in the solenoid for some time, even after it has been short-
circuited and disconnected from the external current source. If we use
a conducting cylinder with a cavity (Fig. 9.1b) in place of a solenoid,
the current will flow for an even longer time than in the first case,
since the resistance of the cylinder is lower than that of the solenoid.
It happens that if a closed-circuit current flows within a neutron star
(Fig. 9.1c), the current can flow for many millions of years, or even
longer, if the matter of the neutron star is superconducting.

"The question arises, 'How is current generated in a neutron star,
and what is the size of this current?'

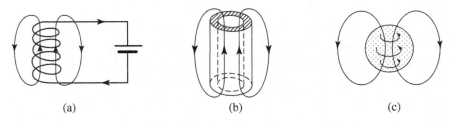

(a)   (b)   (c)

Fig. 9.1 If a solenoid (a) is short-circuited, the current continues flowing for a short
time. This time duration is longer in the case of a conducting cylinder (b) and is much
longer again in the case of a neutron star (c).

"To answer this question, we need to know how neutron stars are born. At present, it is thought that neutron stars are products of the evolution of a certain class of stars: not too large and not too small. Such stars at first, like the Sun, burn hydrogen into helium, and then they fuse heavier nuclei. This process goes on until the core of the star is composed of Fe nuclei and electrons. The fusion of Fe nuclei into still heavier nuclei is energetically unfavorable. Fusion reactions stop and, since the release of energy which could lead to expansion has stopped, the star begins to contract. Compression of the star leads to an increase in the energy of its electrons until at some point the fusion of protons with electrons to form neutrons becomes energetically favorable. Excess energy is taken away by neutrinos — particles that easily pierce the whole thickness of the star. Roughly speaking, the reaction cools down the matter from which the star is made and, under the force of gravity, the matter drops to the center of the star. A monumental event occurs: most of the matter of the star implodes into a very compact blob whilst the star's shell explodes outwards at very high speed. This phenomenon is observed as a supernova.

"From time to time supernovae occur in our Galaxy as well as in other galaxies. For example, a particularly bright one was seen on February 24, 1982. Another famous supernova remnant is located in the Crab Nebula. According to Chinese manuscripts, it first appeared in 1054 and for some time shone like the Moon. The Crab Nebula is actually a sparkling cloud of plasma — a mixture of electrons and ions. This cloud emits light because electromagnetic waves are emitted when electrons are accelerated to very high energies. These electrons are called ultrarelativistic electrons. The electrons produce this radiation as a result of the curvature of their trajectories in the magnetic fields which are generated by the cloud.

"Modern astrophysics can explain many observed phenomena through the experience and knowledge accumulated by physicists using numerous Earth and satellite-based facilities. In particular, the physics of supernova explosions and the properties of the remnants

generated after such explosions can be understood. The Crab Nebula
had been the subject of study even before neutron stars (pulsars)
were discovered. Soon after the discovery of the first pulsar it was
assumed that pulsars were neutron stars, and the search for a pulsar
in the center of the Crab Nebula began. The search was a success: in
1968 American scientists discovered a periodic source of radiowaves
in the center of the Crab nebula. It is interesting to note that the
radio source coincided with the optical radiation source, i.e. with the
star close to the center of the Crab Nebula. The Crab Nebula pulsar period is 0.033 seconds: the rotation velocity of the appropriate
neutron star is very high.

"Thus, a neutron star is born following a supernova, when the
central part of the star implodes to very small dimensions. Now we
can examine how the neutron star can possess a magnetic field.

"We have to recognize that even ordinary stars have magnetic
fields. For instance, the Sun has spots with a magnetic field of several
kiloGauss, and has an average magnetic field of several Gauss. If we
remember that the Earth's magnetic field is of the order of half a
Gauss, then it is clear that the magnetic field at the Sun is relatively
low. However, there are ordinary stars with a magnetic field of the
order of several kiloGauss. What would occur if such a star was
suddenly compressed to the dimensions of a neutron star?

"Let us return to Fig. 9.1. It shows schematically how the magnetic field of a star is generated by ring current. This current is very
similar to the ring current of the conducting cylinder, and to simplify
our task we assume that this premise is true.

"Let B_0 be the average value of the magnetic field inside a tube
conductor (Fig. 9.1b). Let us imagine that this is a superconductor and that the current flows eternally. It is easy to see that inside
this closed conductor one cannot change the magnetic flux. Indeed,
as Faraday proved, the change of a magnetic field inside a closed
circuit leads to the generation of an electric field in such a circuit. Here, however, we have a superconductor, so the electric field
inside it is always zero: any traces of an electric field are immediately

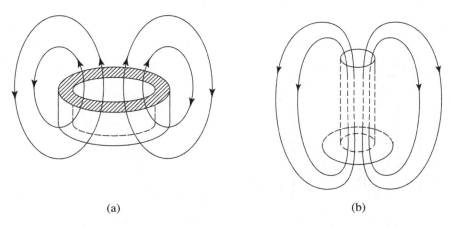

Fig. 9.2   If an ideal conductor ring with current (a) is compressed radially (b) then the magnetic field will increase due to magnetic flux conservation.

eliminated by the appropriate displacement of the superconducting electrons.

"Let us now assume that the conductor (Fig. 9.1b) is not very rigid and can be deformed as shown in Fig. 9.2a,b.

"Suppose that the current-carrying ring (Fig. 9.2a) is compressed radially into a tube (Fig. 9.2b). The magnetic flux is conserved so that the magnetic field in the tube becomes much higher than in the original state. Let B_0 and S_0 be the initial values of the magnetic field and the internal cross-section of the ring, respectively; and B, S be the corresponding values for the tube. The magnetic flux, which is equal to the product of the magnetic field and the cross-sectional area, remains constant. Therefore the magnetic field value in the final state is equal to:

$$B = B_0 S_0 / S$$

If S is much lower than S_0, the magnetic field increases many times compared to its initial value B_0."

"And in this way one can increase the magnetic field?" Sasha asked, very surprised.

"Yes, it is possible," said L.A. "Such a method was invented by the famous Russian scientist Andrei D. Sakharov. He developed a

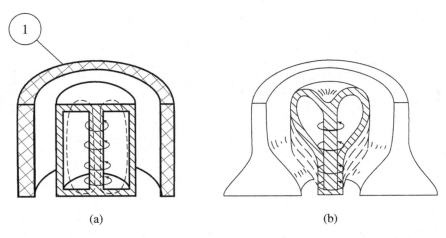

Fig. 9.3   A. D. Sakharov's idea to produce a very strong magnetic field with the help of explosive compression: (a) a current-carrying shell surrounded by an explosive jacket (1); (b) the detonation wave propagates through the explosive and compresses the current-carrying shell.

number of schemes for explosive magnetic generators based on this fundamental idea. Their principle of operation is shown in Fig. 9.3a,b.

"It is convenient to begin with a ring solenoid so that the magnetic lines do not go far from the area inside the conducting shell. In Fig. 9.3a each coil of such a solenoid goes first along the axis and then goes back to the starting point along the periphery. As a result, a multi-coil shell with axial symmetry is obtained. Then explosives are placed around the cylinder. The solenoid is connected to an external current source, and an explosion is initiated from the bottom of the cylinder. First the explosion will short-circuit the coils at the bottom of the solenoid. Then the detonation wave will start moving upwards, pressing the coils and shifting their short-circuit point upwards. The solenoid cross-section will reduce and the magnetic field will increase. Finally the explosion will reach the upper side of the solenoid and the magnetic field will reach its maximum. If the explosive does not surround the whole length of the solenoid, then a cavity with a very strong magnetic field will be formed at one end. As the explosion proceeds very quickly, even ordinary copper shells can be considered during that time as superconducting.

"These 'explosively pumped flux compression generators' allow magnetic fields of up to 25 megaGauss to be produced.

"Now let us return to stars. When a star contracts to form a neutron star the magnetic flux is conserved. The magnetic field thus increases proportionally to the decrease of the star's radius squared. Above, we found the dimensions of a neutron star to be a factor of 10^5 smaller than those of an ordinary star. Thus, the magnetic field of a neutron star should be 10^{10} times higher than that of the original star, prior to its collapse. If the original star, like the Sun, had a field of several Gauss, the magnetic field of the neutron star would reach 10^{11} Gauss; if the initial field was of the order of several kiloGauss, the final value of the field would reach 10^{13}–10^{14} Gauss.

"Thus, a neutron star naturally assumes a magnetic field of 10^{11}–10^{13} Gauss, and, in extreme conditions, up to 10^{14} Gauss.

"The collapse of the core of a heavy star into a neutron star occurs very quickly, so the magnetic field could be considered as being 'frozen' into the matter, i.e. the superconductivity approximation will suffice. After compression, the magnetic field can steadily decay due to the finite electrical conductivity. Estimations show that the field decays over several million years.

"Thus, a neutron star has a superstrong magnetic field, which significantly affects the properties of matter. Additionally, the rotation speed of a pulsar is very high, so a number of very interesting and unusual processes occur on its surface."

10

Physics of the Pulsar

"I'll try to tell you as much as possible about pulsars," said L.A. "Almost as much as I know myself. In doing so, I will present to you my own understanding, which may not necessarily coincide with the opinions of my colleagues who are more experienced in this area. I am not a specialist in astrophysics.

"So, let us begin with the most important property of pulsars, thanks to which they got their name, and this is periodic radioemission. Pulsar is simply a contraction of 'pulsating star,' i.e. the star which emits pulsed radio signals. It was thanks to these radio signals that pulsars were discovered. Later it was found that radiation bursts are observed in all regions of the electromagnetic spectrum: periodically repeated bursts are seen not only in the radio range, but also in the optical and X-ray ranges.

"The most amazing thing is that these bursts are repeated with a surprisingly precise period: it seems that a great number of cosmic watches are scattered throughout the universe. However, in all the cases where accurate measurements have been made, it has been found that the periods of pulsars slowly increase over time. If we denote a pulsar's period by P, the rate of change of the period with time is $\dot{P} \sim 10^{-15}$ sec/sec. The most noticeable irregular changes of the periods are sudden increases in the rotation velocities, called 'glitches'.

"Over a thousand pulsars have been discovered, and their periods range from 1.5 milliseconds up to several seconds. This speaks to the fact that pulsars are compact objects. In 1.5 milliseconds, light can only travel a distance of 450 km, and the speed of light is the maximum speed at which signals can be transmitted i.e. 450 km is the distance over which a causal relationship can act. Thus, a pulsar as a source of signals should be very small — not more than hundreds of kilometers in diameter.

"The very high frequency of the signals and the small size of the emitter led to the opinion that a pulsar is a swiftly rotating neutron star. Its diameter should not exceed 30 km, even though its mass is comparable to the mass of the Sun. The frequency of the signals is related to the inhomogeneity of the emitting surface, and to the fast rotation of the star together with the emitting spot — a kind of antenna."

"A pulsar is a neutron star having dimensions of order 10^5 smaller than those of the Sun. Since its most probable origin is as the collapsed core of a star that has gone supernova, it is expected that a neutron star will possess a very high magnetic field. With an ordinary star, such as the Sun, the magnetic field is of the scale of 1–2 Gauss. The magnetic field of massive stars can reach several hundred Gauss and even up to tens of kiloGauss. Since the magnetic flux is conserved during collapse, when the star is compressed by a factor of 10^5, the magnetic field will increase by a factor of 10^{10}. If, for example, the initial field of a star before compression was 100 Gauss, then the resulting neutron star will have a field of 10^{12} Gauss.

"Thus, from the point of view of astronomical distance scales, a pulsar is a 'tiny' neutron star which rotates with a very high velocity and which possesses a superstrong magnetic field at its surface. The soft X-ray radiation of the pulsar tells us that the temperature of its surface reaches several million degrees, i.e. three orders of magnitude higher than the Sun's surface temperature. A pulsar is quite a strange object, isn't it?

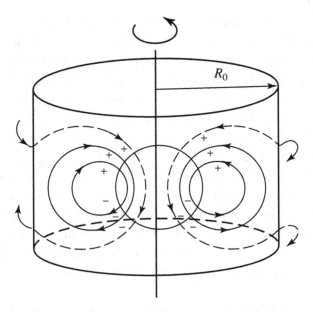

Fig. 10.1   A rotating neutron star carries the magnetic field lines inside a cylinder of the radius R_0 which rotates at the speed of light. The magnetic field lines outside of this cylinder spiral out.

"Now let us try to understand what occurs outside the pulsar. A pulsar is like a quickly rotating magnetic dipole. Let us assume for simplicity that the pulsar's axis of rotation coincides with its dipole axis (Fig. 10.1).

"Let R_0 be the radius of the light cylinder, i.e. the cylinder which, co-rotating with the neutron star, has a rotation velocity $v = \omega R_0 = c$, where c is the speed of light, $c = 3 \cdot 10^8$ m/sec. Here $\omega = 2\pi/P$ is the pulsar's angular velocity, where P is the pulsar's period, which is assumed to coincide with the period of its rotation. Thus, $R_0 = c/\omega \simeq 5 \cdot 10^4 P$ km. If, for example, P is 10 msec, then the radius R_0 is equal to 500 km. Let us now mark the area where the magnetic field lines are completely inside the light cone (see the cylinder of radius R_0 in Fig. 10.1).

"The corresponding volume could be called the pulsar's magneto-sphere — it rotates with the pulsar. To understand why this happens, let's examine a charged particle located within this area. Let this particle be at distance R from the rotation axis, and, to simplify things, let us assume that it is in the equatorial plane $z = 0$. If this particle rotates with velocity v along the circumference, then it experiences a Lorentz force $F_B = eBv$ directed radially. Suppose there is an additional electric field E, then due to the symmetry in the plane $z = 0$ it can only be directed radially so that the total force is equal to:

$$F = eBv + eE$$

If the particle rotates together with the magnetic field lines then this force should be equal to zero.

"Conversely, if an electric field exists that compensates the Lorentz force in the marked area, then all the area rotates as a rigid body together with its charged particles and magnetic field lines. Since the Lorentz force is directed perpendicularly to the magnetic field, the electric field is also perpendicular to the magnetic field lines. In other words, the magnetic field lines inside the magnetosphere are equipotential lines because the longitudinal component of the electric field is equal to zero. For an electric field to exist, the magnetosphere must be filled with charged particles, i.e. with plasma. The pulsar's surface should also have charges — negative closer to the equatorial plane, and positive closer to the poles. The electric fields on a pulsar reach fantastic magnitudes since the rotation velocity and the magnetic field values are extremely high, but inside the magnetosphere this does not significantly affect particle motion — along the magnetic field lines the forces have no effect on the charged particles because the longitudinal E-component is zero.

"The full picture changes drastically outside the magnetosphere. The field lines cannot be closed in this region — they pierce the light cylinder and curve outwards like the jets of a Segner-wheel. This means that outside the magnetosphere some radial component of the electric field appears that pulls the charged particles from the surface and accelerates them to high energies. Flying away from the light cone, they produce a very powerful radiosignal which could be detected by antennae on Earth.

"If the dipole axis does not coincide with the pulsar rotation axis, the whole picture becomes more complex. However, let us go back to the pulsar surface to try to understand what a neutron star looks like.

"It turns out that a neutron star is somewhat similar to the Earth (Fig. 10.2). If we do not go into too much detail, we can say that the pulsar has a solid crust (2) with an atmosphere above it (1), and nuclear matter layers below it. It is obvious that the very upper layer (1) is the lightest one. If it is a gaseous layer, then it is the analog of the Earth's atmosphere. Next we reach a solid shell, the density of which increases as the center is neared. This shell is primarily made of Fe-nuclei which float in a 'solution' of electrons and neutrons. Since the matter is heavily compressed, it should be solid, i.e. crystalline. As we go inside the pulsar the matter will look more and more like heavily compressed nuclear matter — like the nuclei of atoms. Now let us think a little about the upper layers of the pulsar.

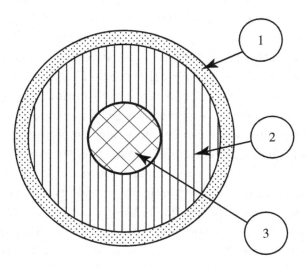

Fig. 10.2 Schematic structure of a neutron star: an atmosphere (1) surrounds a solid crust (2) beneath which lies a high density core (3). [Figure not to scale.]

"First of all, let us estimate the value of g — the acceleration of free fall on the pulsar's surface. We can do it in the following manner. First, let us find the Earth's acceleration during its rotation round the Sun:

$$g_E = \omega_E^2 R_E = \left(\frac{2\pi}{T}\right)^2 R_E$$

Here g_E is the acceleration of the Earth towards the Sun, $T \simeq 3 \cdot 10^7$ sec is the period of the Earth's rotation, i.e. one year, and $R_E \simeq 150 \cdot 10^6$ km is the distance from the Earth to the Sun. We find $g_E \simeq 6 \cdot 10^{-3}$ cm sec^{-2}.

"The radius of a pulsar, $r_0 \simeq 15$ km, is 10^7 times smaller than the radius R_E of the Earth's orbit; and the period of a pulsar's rotation is $\sim 10^{-10}$ years. Hence, for a pulsar with the same mass as the Sun, the surface acceleration g should be 10^{14} times higher than g_E. It may be that the pulsar mass is somewhat less than the mass of the Sun. Let us accept for g the value

$$g \simeq 6 \cdot 10^{11} \text{ cm sec}^{-2}$$

This value is $6 \cdot 10^8$ times higher than the acceleration due to gravity on the Earth's surface.

"Now we can estimate the atmosphere thickness on the pulsar by analogy with the Earth. On Earth the atmosphere is about 10 km thick, i.e. 10^6 cm. It would be $6 \cdot 10^8$ times thinner at the same temperature on the pulsar. Since the pulsar temperature is 4 orders of magnitude higher than room temperature, then, using Charles's law, the thickness of the atmosphere on the pulsar is 10 cm. It is low in its absolute value, but it is $\sim 10^{-3}$ times the pulsar's radius — like the atmosphere compared to the Earth.

"Now let us examine the crust. It is solid matter to a depth of 2–3 km and consists of crystals made of a mixture of Fe-nuclei, neutrons and electrons. Such a crust may be similar to the Earth's core. If the crust has not attained complete homogeneity and equilibrium stratification, gravity and the magnetic field may cause cracks to occur leading to 'starquakes'. These starquakes are one of the reasons for increases in pulsar rotation rates. (However, there are hypotheses that claim that the starquakes are caused by a redistribution of angular momentum in deeper layers where the nuclear matter is in a state of superfluidity). Such cracks can, as on Earth, cause the formation of geysers and volcanoes.

"Finally, let us discuss what can occur in the upper layer of solid crust, i.e. on the pulsar's surface. One should take into account the presence of the superstrong magnetic field, which can drastically change the shape of atoms. At a field of 10^{12} Gauss, i.e. three orders of magnitude higher than the critical value B_0, atoms are strongly elongated along the magnetic field. The positively charged nuclei are exposed, and can be pulled towards the electron clouds of adjacent atoms. Polymer structures are formed. Atomic bonds in such structures can be three orders of magnitude stronger than in similar structures on Earth. This means that rigid structures can exist on a pulsar even at temperatures of millions of degrees.

"Thus, in the uppermost layers one can imagine the existence of a gaseous atmosphere and thin layers of liquid, covering a solid crust formed from elements with atomic weights similar to that of iron.

"One may ask how uneven the pulsar's surface can be: are there any mountains, plains or valleys? To answer this question we may again use the similarity with the Earth. On Earth, mountains are of the order of kilometers high, whereas the scale of fragile objects — trees, for example — is tens of meters. On the pulsar the acceleration due to gravity is $6 \cdot 10^8$ times higher, but the strength of materials is three orders of magnitude greater. Therefore, the inherent dimensions of corresponding structures should be at least 10^5 times smaller: 'mountains' should be centimeters high and more fragile structures, such as trees, shouldn't be taller than one millimeter.

"These small structures can still significantly affect the ejection of charged particles from the pulsar's surface. In places with slight roughness and sharp but small peaks such ejections become much easier. Consequently, the ejection of high energy particles into the upper layers of the pulsar's magnetic field can be rather uneven;

conventionally speaking, 'hot spots' can exist with better conditions for the release of charged particle beams.

"Thus, the pulsar radiation can be determined by an interplay of surface and volumetric processes in the pulsar's plasma."

"L.A., your story was fascinating. I clearly saw small mountains, seas, rivers and plains, but what about life on the pulsar? Maybe numerous herds of tiny elephants and hippopotamuses rove over the plains, and small monkeys the size of mosquitoes climb grass-like palm trees." asked Misha.

"Now here we have a completely false supposition," said L.A. earnestly. "However, the question of how far the pulsar's development can go is not a groundless one. What development means and what life is are questions for another day."

And L.A. invited the boys to visit him in a week.

11

Non-linearity and Self-organization

"The question, 'What is life?' is a very complex one," L.A. began his next lecture.

"We will approach this question gradually, step by step, starting from the simplest things. The simplest approach to the explanation of the Universe is based on the statement that the Universe was created by God. I will not go into detail now discussing the role of God. However, there is no reason to think that God started the creation of the Universe by violating His own laws. It is more sensible to assume that the Universe, no matter who created it, is developing in accordance with certain laws. This is why the development of the Universe should be studied on the basis of the assumptions that we have at present, and on the basis of the knowledge of those laws that are considered to be proved in practice.

"The Universe, as we see it now, was created through a long process of evolution. First simple structures were formed, and these gradually turned into more complex ones. Finally there resulted that unique Universe that is praised by poets and artists, and by us.

"The simplest ways of forming new structures are well described by the laws of physics. The formation and further development of such structures is called self-organization. In order to understand how these processes occur, we will have to start with even simpler things. To begin with, we will become acquainted with 'non-linear' phenomena, but to understand non-linearity, it is first necessary to define what linearity is.

"Linear processes are processes in which all the characteristic physical variables are related to each other by linear expressions. The simplest example of a linear process is that of the harmonic oscillations of a point mass under the effect of a quasi-elastic force.

"Let x be the coordinates of a particle. A quasi-elastic force has a strength which is proportional to the deviation from equilibrium:

$$F = -kx, \quad k = \text{const}$$

The potential energy of a particle under the effect of this force looks like a parabola (Fig. 11.1). If there is no friction, then a point mass driven out of equilibrium undergoes simple harmonic motion: its coordinates vary with the cosine or sine of time. If friction exists, the oscillation amplitude decreases with time until the particle comes to rest at the point $x = 0$, which corresponds to the minimum of potential energy.

"If an external force F_0 is applied to the point mass, the equilibrium is shifted to the point $x_0 = F_0/k$ (see Fig. 11.1). As we see, the coordinate x_0 is related to F_0 by a simple linear relationship. Therefore, if a particle is affected by several forces, then the coordinate of the particle's equilibrium is simply proportional to the sum of those forces. One can say that a principle of superposition governs

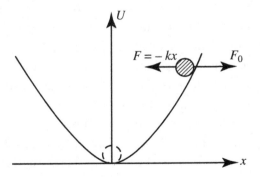

Fig. 11.1 Particle in a parabolic potential well produced by a quasi-elastic force $F = -kx$.

the displacement: forces do not interfere with each other, and the elementary displacements produced by different force components are in no way affected by each other.

"A more complex linear phenomenon is the propagation of small amplitude waves in a quasi-elastic medium. The harmonic waves in such media have sinusoidally varying amplitudes that are functions of both position and time. The principle of superposition is also true for linear waves. Different waves do not interfere with each other, and each one propagates as a single wave. For example, this is normally true for sound and the low amplitude light to which we are accustomed in our everyday life.

"Non-linear phenomena are much more complicated. The ancient Greeks knew this intuitively, which is why one of the most vivid and important examples of non-linear effects was first debated by Greek philosophers. I have in mind the well-known paradox of Buridan's ass: let an ass stand equidistant from two haystacks. The ass is hungry, but since it is standing exactly between two haystacks it has no particular reason to prefer one haystack to another. Therefore, it will die of hunger having not started its meal.

"Every non-philosopher knows that this isn't true: a hungry ass would begin with one haystack and then go to the other. Where is the paradox? The initial state of the ass between the two haystacks is unstable, which is why it cannot last forever. A small perturbation is enough for the ass to turn either to the right or to the left. For the observer it is not important whether the ass has made its choice as a result of e.g. some breath of wind which carried the scent of hay, or has made a conscious decision. In either case the initial symmetry will have been violated. The destruction of the initial symmetry is called spontaneous symmetry breaking.

"This process can be illustrated by a simple physical example. Let us examine a point mass subject to forces such that the potential energy has two minima (Fig. 11.2). Let the potential energy curve be symmetrical and the mass point be on the top of the potential hump at $x = 0$. This state is unstable: a small initial perturbation

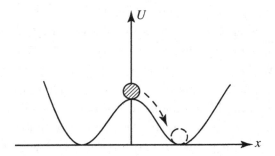

Fig. 11.2   Non-harmonic two minima potential well and the corresponding bifurcation.

is enough for the particle to slide down to one of its minima. Due to friction, the particle will settle at the bottom of the potential well after several oscillations. This is a physical analogue of Buridan's ass's breakfast.

"In its final state, the particle will be either at the right or the left of the minimum of the potential energy (Fig. 11.2). A spontaneous breakdown of symmetry has occurred. As is seen from Fig. 11.2, the potential energy does not look like a parabola and the force affecting the particle is not linear. On the contrary, spontaneous symmetry breaking is significantly non-linear. The result of this non-linear process is the reaching of a new state which is strongly different from the initial one, even from the point of view of such a fundamental property as symmetry.

"Spontaneous symmetry breaking occurs in a large class of natural non-linear phenomena. A drastic example of this took place at the beginning of the Universe's evolution. As we know, the Universe consists of a great number of particles — neutrons, protons, electrons — but not antiparticles — antineutrons, antiprotons, positrons. Nothing prevented the Universe from taking the other option: the choice was made analogously to Fig. 11.2, by the spontaneous breaking of symmetry.

"I can give you a number of examples of the spontaneous breaking of symmetry. For example, every domain of a ferromagnet is in a state of minimum energy when magnetized. However, the direction of the magnetization is chosen by the spontaneous breakdown of symmetry. Another example is the magnetic field of the Earth or a pulsar: assuming that the magnetic moment is set along the rotation axis, then the direction of magnetization, whether along or opposite to the direction of the angular momentum vector, is chosen spontaneously.

"Now let us study in detail the transition from a state (Fig. 11.1) with one equilibrium point to a state (Fig. 11.2) with two positions of equilibrium. To do so, let us choose the potential energy in such a way that it can allow the presence of a 'hump' in the background of a regular parabola (which corresponds to a quasi-elastic force). For instance, let the potential energy be:

$$U(x) = x^2 + P^2 (1 + x^2)^{-1}$$

The second component describes the potential 'hillock' whilst the first component describes the potential 'well'. When changing P, one can also change the form of the potential well, or, in other words, control the potential energy curve. Therefore P is called a control variable. The function of the potential energy, given above, can be rewritten as follows:

$$U(x) = \left(\sqrt{1 + x^2} - \frac{P}{\sqrt{1 + x^2}} \right)^2 + 2P - 1$$

If $P < 1$, this function has only one minimum at the point $x = 0$. If $P > 1$, then the function $U(x)$ has two minima: one at each of the points where the expression in brackets becomes zero. Thus, in this case, the potential energy minima are solutions to:

$$1 + x^2 = P$$

Sometimes the quantity $P_c = P - 1$ is introduced so that when $P_c < 0$ there is only one minimum whereas when $P_c > 0$ there are two. The variable P_c is called the 'order parameter', and the point $P_c = 0$ is usually called the bifurcation point, i.e. the point with 'two forks'.

"There is a repeated episode in Russian folk tales which represents a typical example of bifurcation. A Russian bogatyr (a hero of

Russian folklore) is riding a horse and approaches a fork in the road. There is a huge stone in the road which says 'You will lose your horse if you go to the right and you will lose your head if you go to the left'. The bogatyr, being fond of his horse, chooses the left road and defeats the dragon.

"Now we know that bifurcation is typically a non-linear process, but Buridan's ass solved the complex puzzle of non-linear bifurcation without thinking. Bifurcation in physics is a very common phenomenon and it can be solved effectively.

"As an order parameter, various values could be considered. If, for example, 'ordering' occurs when the temperature T goes down, as in the case of ferromagnetic materials, then the difference $T - T_c$ where T_c is the temperature of ordering, is the latent ordering parameter. The value and the direction of the magnetic moments are the obvious parameters of order.

"In non-equilibrium systems with complex structures, the parameters of order can either be given arbitrarily or created by certain parts of the structure with respect to other parts.

"Even the simplest state, shown in Fig. 11.2 is non-equilibrium from a thermodynamical point of view. I will now explain the essence of this to you. There is a special branch of physics called thermodynamics. It is the study of processes in conditions of thermal equilibrium. This means not only that all parts of the system under study have the same temperature, but also that those parameters that can take one of several values have an equal chance of taking each one. If the particle (Fig. 11.2) has two positions of equilibrium on the microscopic scale, then due to thermal motion it can jump from one well into the other from time to time. Therefore, on average no violation of symmetry will occur. With 1/2 probability, the particle will be present either in the left or in the right well. However, if the temperature isn't high enough, the time between jumps from one minimum to the other can be enormous.

"In such a case, the system shown in Fig. 11.2 can be considered as a cell of memory: a particle will stay forever in the half of the potential into which it is initially placed. This is an elementary cell of memory with only two positions. Such a cell is said to store one 'bit' of information; and such a cell is non-equilibrium from a thermodynamical point of view.

"If we create a number of identical cells of information, then the amount of the information stored will grow. If we have N cells, then the memory can store N bits of information.

"It can easily be seen that such a system has 2^N different states, and the probability of a single state is equal to $p_N = 1/2^N = 2^{-N}$. Therefore, the total amount of information I in a given state is:

$$I = N = -\log_2 p_N$$

Here $\log_2$ is the logarithm with base 2.

"The greater the number of memory cells, the larger the volume of information that can be stored. Consequently, the probability of a given single state decreases as the volume of information stored increases. When speaking of the volume of information, we simply have in mind the number of bits, irrespective of the content of such information.

"For example, the text on this page contains some amount of semantic information which could be placed in the memory of a computer, or copied onto a disk. In this case we would have a 'compressed' version of semantic information. However, we can approach this problem in a simpler way. Let us break the page into lines with a large number of dots in each line. This is how it is done in television. Now we can darken those dots which correspond to letters, and leave the other dots, which are in between the letters and punctuation marks, white. We will again have a memory with a great number of cells where the text is the information stored in the memory. The number of bits that can be stored on one page depends on the number of dots on the page. In other words, the greater the number of dots, the better the resolution of the text, and, consequently, the larger the volume of information. In this case the text is memorized not by its logic, but simply as a picture.

"Let each letter be composed of a number of tiny dots of the same size. Let the blank area be broken into separate dots with the total number of dots, black and white, equal to N. We can assume that each dot can remain in two states — black or white. That is how information displays are arranged in airports. It is obvious that the total number of states of dots on this board is equal to 2^N, and the probability of a certain text, i.e. picture, being displayed is equal to 2^{-N}, i.e. one state per 2^N possibilities. If N is high enough, i.e. the resolution is good enough, the probability of any given state is very low.

"If we assume now that these dots are small enough to be affected by thermal fluctuations, sooner or later the color of the dot will change. It is clear that the picture will become vague over time, and will eventually converge into an even gray background. This final state corresponds to thermodynamic equilibrium. It is also clear that we will never obtain text out of this gray background through random fluctuations. This phenomenon is called irreversibility.

"We are accustomed to irreversibility in our every day lives. 'Life can never be reversed,' as one popular song states. This is true. Everything that happens around us happens in an irreversible way.

"The famous English physicist Stephen Hawking likes to show a simple video. A cup of coffee rests on a table and then slides to the edge, falls down on the floor and breaks up into many fragments. The coffee is spilled on the floor. 'We are used to such a course of events,' says Hawking.

"Then he plays the video backwards. The porcelain fragments join each other, producing a cup, and the spilled coffee flows into the cup. Then the cup 'jumps' onto the table and rests there.

" 'No one observes this course of events,' says Hawking.

"This is an example of irreversibility. It seems that this is the way things are: highly organized states are transformed into less organized ones. However, things are not as simple as they seem. We started with a cup of coffee on a table, but several years ago there was no such cup. A process existed that led to the creation of a cup from clay. This process was also irreversible, but it created a piece of art from amorphous clay; a fine and complex structure from a homogeneous lump. We cannot imagine this process existing without man. But man himself is the result of evolution, and this evolution did not violate the principle of irreversibility. It seems that there is irreversibility, and then there is irreversibility. Let us try to understand this in more detail.

"Let's go back to our information board. Let $\Gamma = 2^N$ be the total number of states consisting of N dots that can be either black or white. If the initial text corresponds to one given state, then as the text dissipates due to thermal fluctuations the number of possible states increases. In its limit, it tends to $\Gamma = 2^N$, when all equally probable states are realized in time as a result of thermal fluctuations.

"In statistical physics, which is a branch of physics like thermodynamics, there exists a concept of 'entropy'. The entropy of a state is determined by Boltzmann's well-known relationship:

$$S = k \log \Gamma$$

Here S is the entropy, k is Boltzmann's constant, log is the natural logarithm, and Γ is the number of possible realizations of a given macroscopic state. Boltzmann's constant appears here only because we conventionally measure temperature in degrees, and the expression ST has the dimensions of energy. If we measure temperature in units of energy, then Boltzmann's constant can be omitted. Therefore we can simply write down:

$$S^* = \log \Gamma$$

The $*$ index shows that the temperature is measured in units of energy.

"Thus, entropy is equal to the logarithm of the number of different microscopic states which belong to a single macroscopic state. If we begin with only one state on the information board, then the entropy is zero. As the picture loses its clarity and the number of possible realizations tends to $\Gamma = 2^N$, the entropy grows and reaches its maximum. Therefore, the irreversibility of a closed system corresponds to the growth of entropy, and as the entropy grows information is destroyed. Consequently, there should be a relationship between entropy and information.

"In order to simplify the quantitative relationship between information and entropy, we had better amend our definition of information slightly. Above, we agreed to measure information in bits, so that the total amount of information that could be placed in N memory cells was equal to N bits. However, we don't necessarily need to express information in bits. For example, we measure computer memory in terabytes where each byte is equal to eight bits. To establish the connection between entropy and information, it is convenient to use a unit for information that differs slightly from the bit. Let us instead of taking a logarithm with base two, take a natural logarithm. Then for an information board with N dots and with 2^N total states, the information of a certain selection is defined as:

$$I^* = \log(2^N) = N \log 2$$

As we see, the new definition of information is $I^* = I \log 2$, where I is the number of bits. This new definition was introduced by Shannon.

"We are now ready to establish a relationship between information I^* and entropy S^*. Let us perform a thought experiment. Let some text be displayed on a board with N cells. The volume of its information is:

$$I^* = N \log 2$$

The entropy of such a state, which has no fluctuations, is equal to zero: $S^* = \log 1 = 0$. Now let thermal fluctuations begin in a spot with n cells. Then from the whole area of the board, the information in the spot with n cells seems to have been removed. On average this spot is of gray color and cannot store any information. Thus, the information on such a 'damaged' board is equal to

$$I^* = (N - n) \log 2$$

The entropy of the 'damaged' board grows: the number of states that undergo thermal fluctuations is equal to $\Gamma = 2^n$. Thus,

$$S^* = \log \Gamma = n \log 2$$

As we see, the following relationship holds:

$$S^* + I^* = \text{const}$$

The volume of information decreases proportionally to the growth of entropy S^*.

"It is proved in thermodynamics that the entropy of a closed system grows steadily until it reaches its maximum. This is how irreversibility can be expressed in physical units. If we have some structured physical system, for instance a cup of coffee, then it possesses a high degree of order which, in its turn, can be expressed in information units. The irreversible process — the cup falling and breaking — leads to the destruction of the structure and a decrease in the corresponding information content. The entropy of such a process grows.

"Therefore, units of information are units of 'order', whereas units of entropy are units of 'disorder'. In other words, negative entropy is a measure of order. This is why the French physicist Brillouin introduced the notion of negentropy — entropy with the opposite sign — as a measure of ordering. One may say that the higher the negentropy, the more perfect and ordered the system is. As a system degrades due to irreversibility, the negentropy decreases. This process goes on in every closed system. In other words, the fact that the cup of coffee breaks is quite a natural process and it corresponds to the direction of time in irreversible processes.

"The question again arises of how a cup of coffee could have appeared if all irreversible processes lead to a continuous growth of entropy, i.e. to 'dead' thermal equilibrium? Another question: how could the initial text have appeared on the board if every process is irreversible and irreversibility leads to the destruction of text? The answer is simple — in both cases the orderly structures appear from the outside: somewhere outside the limits of the system under consideration a nonequilibrium process is involved.

"In order to describe these processes, physicists introduce the notion of an open system. An open system can exchange information, as well as energy, with external media. It seems as if the system acquires a supply of negentropy from outside in order to avoid the irreversible approach to equilibrium.

"The notion of an open system answers the above questions. However, it raises another one. If we join the external media to the open system we again have a closed system in which entropy can only grow. We have a paradox.

"Fortunately, it only seems to be a paradox. To solve it, we can consider the following analogy. You will surely agree that water always flows downhill. In ancient times it was known that an irrigation system could be built by attaching a pump to a water mill in place of millstones. Then a tiny part of the river's flow could be pumped to a great height. The main part of the flow tended downwards while some part of the flow was raised.

"A similar process can take place in the transition of a non-equilibrium system to the state with maximal entropy. The entropy grows in the system as a whole, but in some small parts of the system the entropy can decrease. These parts of the system are open systems, they are systems with self-organization, i.e. with growing order. Inside such a system the energy and negentropy may increase, and the entropy decrease. Such a system should eject any excess energy and entropy produced. In other words, this open system should consume energy sources with low specific entropy, i.e. with a high degree of order, and discard 'waste'. Only by these means can an open system sustain its orderly structure or even make it more complex. As you see, we are gradually getting closer and closer to understanding the process of life. However, let me introduce another simple example to you.

Fig. 11.3   The candle flame as a self-organized open system: the candle body and oxygen gas flow towards the flame, and waste CO_2 is taken away by the convective air flow.

"Figure 11.3 shows a candle flame, which can be considered as an open system. The flame is in a steady state as the candle body is consumed. A flame is a self-organizing system whose dimensions are adjusted so as to coordinate with the rate of CO_2 production in the flame. This is an example of a non-equilibrium process with increasing entropy. However, the flame itself is not being irreversibly destroyed. To the contrary, it is being sustained in a steady state with continual replenishment of the matter it consists of. A flame is a simplified model of a creature which needs to eat and breathe to live.

"On the other hand, we can use a simplified model (see Fig. 11.3) which describes bifurcation for the flame of a candle. Actually, a candle has two steady states: burning and extinct. The flame of a burning candle is relatively stable against small perturbations, e.g. breezes. However, a strong air flow will be enough to blow out a candle. To light the candle, we have to burn it again, otherwise it would remain in an extinct state forever. Thus a flame, like the ball in the potential well (see Fig. 11.2) has only two steady states. Hence, it can be considered as an elementary memory cell which can store one bit of information. In one of the states, the burning one, the candle is in an excited self-organized state. Thus, the flame of the candle can be considered as an open self-organizing system which maintains its internal structure due to an external energy supply and its 'metabolism'.

"All complex open systems are created in the same way. Their general scheme is presented in Fig. 11.3. As we see, an open system consumes energy and matter from outside. The consumed energy has low specific entropy content. The open system's internal structure is sustained by the energy and negentropy flows through it. The negentropy consumed by the system is turned into entropy and is thrown outside as heat. The structural matter that is consumed from outside, and which is the negentropy-supplier, undergoes processing and, later, is expelled as waste. The system 'lives' by consuming food which is fresh and rich in calories.

"All complex self-organized systems are created according to the same principles. Take, for instance, the Earth's atmosphere. It is a complex, open, self-organized system. The Sun heats the Earth's surface and the lower atmospheric layers. Vertical air convection warms up the lower layers of air, and leads to a higher rate of heat transfer even to the upper layers. In fact, convection does more than this: it becomes an amplifier of horizontal flows. Weak horizontal wind flows are amplified due to the 'negative viscosity' that arises from vertical convection. A complex pattern of broad-scale flows of cyclones and anticyclones as well as transfer flows between them is formed. As a whole, the atmosphere exhibits complex self-organized motion.

"Broad-scale air flows are accompanied by large water transfers: water evaporates from the ocean surface, condenses into clouds and falls as rain deep inside the continents. As moisture accumulates, it gradually gathers into streams and rivers, which flow back into the seas and oceans. This complex picture of air and water movement is regulated by complicated non-linear ties between the separate components — participants in this complexly organized mechanism.

"This process is a complex one. However, its principles can be represented by a simple picture (see Fig. 11.3).

"The atmosphere starts moving due to the Sun's energy, but the energy supply alone is not sufficient for atmospheric self-organization: the energy needs to come together with a negentropy flow. Sunlight's specific entropy is much lower than that of thermal energy with a temperature of about 300 K. Firstly, the Sun's surface temperature is 20 times higher and, secondly, the light energy of the Sun is strongly isotropic. It is at the expense of the high organization of the Sun's energy that the atmosphere can exist. Vapor transfer and precipitation is the atmosphere's 'metabolism'. The atmosphere assimilates water in the form of vapor and secretes it in the forms of rain, snow and hail. The atmosphere's metabolism is a guarantee of the water supply of the whole Earth's biosphere.

"The biosphere itself is built on the basis of a similar scheme. Its ingredients, i.e. plants and animals, are included in the scheme too. A plant is an open system. It uses the Sun's rays as the energy and negentropy supply. Consuming water, CO_2 and O_2, and a number of other elements and compounds from the soil, plants synthesize complex organic materials.

"Animals cannot directly consume and use the Sun's energy. They gain it indirectly while consuming the biomass accumulated by plants. From the point of view of consuming the Sun's energy, herbivores and carnivores are not very different — they both use the already accumulated and excreted energy and negentropy of the Sun.

"The Earth's biosphere together with the flora and fauna, atmosphere and water resources represents a very complex self-organized open system with a number of non-linear ties between its ingredients. Man is the upper link in this self-consistent picture."

"And why is man needed?" Misha asked suddenly.

L.A. was silent for a moment and then answered.

"Your question is a good one. Indeed, there are two questions: does the development of nature have a goal and, if so, what is man's place in this development?

"You must understand that this is a permanent question. It has been pondered by generation after generation, but it has not been answered yet, which is why I will only make some comments. The simplest answer to your question was given by Indian tribes. Proceeding from the knowledge that somebody feeds on animals and plants, they assumed that something should be living on people's thoughts. People, in their system of thinking, exist only to bear ideas and develop intellect. On the higher planes there exists something that can digest ideas and utilize them for its own development. They call such a superbeing 'Great Eagle'. We will probably never know whether or not this Great Eagle really exists, but the idea that man is not the pinnacle of nature's development seems to be true. I prefer the saying of Soloviev, a Russian philosopher: 'The Universe is not a mechanism. The Universe is an organism and this organism is God'. According to this point of view, the Universe is in the process of development, and the moving force of this development is a spirit, or God, if you like. Man is a component, a certain episode in the greater scene of the Spirit's development."

"But how is the brain of a man arranged? Why can a man think?" asked Sasha.

"The brain is a very complex system," L.A. began. "Physicists have started to develop approaches to understanding it using very simple models. These simplified models of the brain are called 'neural nets'.

"The brain consists of a huge number of neurons — about 10^{10} of them. A neuron is a brain cell that can appear in two states: inhibited or excited. Therefore, a neuron is a kind of memory cell with a capacity of one bit.

"The neurons are connected to each other by a multitude of neuron fibers called 'axons'. By means of these axons they can relay signals to each other and, as a result, the neurons can change from the inhibited state to the excited one and back. The result is a very complex non-linear object.

"A physical neural net is a simplified version of a real, i.e. living, neural network. This neural net may be constructed as a set of numerous elements that have only two stable states. These elements can interact with each other in ways that involve not only their immediate neighbors, but also more distant ones. A neural net resembles the information board that we were speaking about earlier, but having more complex relationships between the elements.

"A simple physical analogy could be built as follows. Let each element be a tiny magnet, or a spin, and let each spin have only two possible directions — positive and negative. Such a spin then resembles an elementary memory cell of one bit. An object consisting of a great number of two-state cells is nothing complex. The complexity appears if the spins interact with each other, and if this interaction is not limited to their nearest neighbors. This interaction can be negative or positive: an interaction may tend to arrange spins parallel or antiparallel to each other.

"Under the effect of these interactions a very complex orderly state appears, which is called a spin glass. It is obvious that the whole system tends to a certain minimum energy. A state with minimum energy is called an 'attractor'.

"A spin system can have quite a number of attractors. Each attractor can be considered as an element of memory, i.e. an attractor which corresponds to a minimum of energy should be stable against small perturbations. Under large perturbations the spin system can jump to another attractor. The picture again resembles the one in Fig. 11.2, but with a multitude of minima.

"A neural net can react very selectively to perturbations applied from the outside. If the perturbation affects a multitude of spins in an orderly manner, then it can very easily shift the system from one attractor to another. In other words, the neural net very selectively responds to an external signal with a well-organized structure.

"Regular background noise simply shifts the system slightly over the equilibrium state; so you see, this system is very resistant to accidental noise. Moreover, if in a spin system the form of interactions between spins can change, for example, under the effect of repeating signals of the same shape; then the states corresponding to attractors will change. Such a system reacts differently to different external stimuli. To obtain something similar to 'thinking' one more step is needed — the system needs to change under the effect of signals it creates itself.

"Certainly, this scheme is very simplified and is still very far from the human mind, but it helps to establish approaches to the question of what thinking is. You, young people, are the ones to resolve these mysteries," L.A. finished.

12

On the Pulsar

"Today we will undertake an unusual journey," said L.A. to Misha and Sasha during their next meeting. "We will travel to the Pulsar."

"Really? But how?" uttered Misha.

"I will now explain it to you in more detail. I will have to start from afar. You might have heard about the everlasting dispute about whether sleep is needed by animals and human beings. The simplest explanation is focused on the fact that the brain needs rest: during the daytime the brain is oversaturated by various impressions, and therefore sleep is necessary to calm down, to process the impressions and to move them to more distant memory cells and therefore be ready to accept new impressions. As you probably noticed yourselves, simple and primitive explanations are often far from the ultimate truth. They can fail when discussing complex phenomena. And the process of thinking is one of the most complex phenomena.

"Like other scientists, I myself support the idea that sleep is vital to a man and is a very important component of the thinking process. Roughly speaking, sleep is a phase of man's communication with the spiritual essence of the World. One may say that the daytime is used by man to communicate with the materialistic part of the

---

*Contrary to the earlier sections of the book having a popular scientific nature, the present section should be regarded as scientific fiction, or even fantastic fiction. It is included in the book only to stimulate young readers to dream and fantasize because this is the only way to come across new ideas and thoughts.

world whereas night-time, with its dreams, to communicate with the spiritual one.

"Obviously, if this kind of communication is an actual fact, one can try to translate it into the language of materialistic forces, though it is not imperative at the beginning. We all know that thought is not a simple product of man's existence.

"Therefore, I suggest you take a position that sleep with dreams is one of the forms of man's cognition of the World — not by means of the senses but by deeper involvement with it. If this is true, then our dreams are not a simple result of the brain's 'purification' of 'layers' of unnecessary impressions, but a certain form of perception of the World. Therefore, it is not strange that there exist prophetic dreams, revelation dreams, premonition dreams, etc. The World is real not only because we can understand it by means of our direct perceptions. The reality has a much deeper meaning and our thinking activity is only a small component of the spiritual activity of the World. Based on these assumptions, several years ago I started research aimed at active perception of the world in our sleep. I won't dwell on all the difficulties of my research but I will show the result. Let us go to the experimental hall."

L.A. together with Misha and Sasha left his office and proceeded to the experimental hall. It was a large room packed with various instruments and apparatus.

Unlike semispherical halls, which were very elegant and clean, this was extremely ugly. All the instruments were connected to each other by chaotically spread wires. It looked as if there had been no time to arrange them in order.

"Please follow me to the armchairs," said L.A.

Only now Misha saw three armchairs with lots of wires around them. "They look like dentist's chairs," he thought.

"Exactly like an electric chair," said Sasha aloud.

"Do not be afraid," said L.A.

He took the middle chair and invited the boys to take the two other chairs.

"You probably know," continued L.A. "that in the East a special ritual of meditation exists. By means of self-hypnosis the man assumes a state in which he feels he can perceive the supreme worlds.

"In this hall we can perform something of the like. Using the instruments we have we can quietly and without any health effects put a man into a meditative state and then bring him back to reality. Moreover, the instruments allow the performance of objective meditation, i.e. meditation to obtain certain information, the nature of which is preset by means of the same instruments. I won't disclose all the possibilities of this equipment, I will just explain the coming experiment.

"I suggest and believe that you will accept my invitation to take a journey to the Pulsar. We will be half lying in the same armchairs. A number of monitors and electrodes from the surrounding instruments will be attached to our bodies. All these instruments together with the supercomputers located in the adjacent rooms form a unified cumulative information system. The most important components of this system are situated outside the building. These are a large optical telescope and a very sensitive radio telescope with a large parabolic mirror. Both of them are oriented on the Pulsar and receive its rays. They are arranged in such a way that they allow reception of not just the rays but also those correlations between quanta of optical and radio ranges that are transported with the rays. It appears that it is just those correlations where the basic information on an object is focused. We cannot translate this information into text that could be read yet. However, if a man is brought into meditation, then his brain directly perceives this correlative information. He would seem to be dreaming and his images are guided by the information he receives in the form of correlative relationships of radiated quanta.

"It looks like a dream induced from the outside. If you agree to participate in this experiment, the three of us would find ourselves on the Pulsar while sleeping, or half sleeping."

"And what do we expect to see there?" asked Sasha.

"It depends on what exists there," answered L.A. "To be more precise, it depends on the degree of organization or, if you wish, 'spirituality' of the Pulsar. If the Pulsar's surface is a simple lifeless desert, then that will be all of our observations. But if a very complicated organization exists there, i.e. something like life, then we could perceive it and try to understand it."

"But what if in this 'life' only the tiniest creatures exist, so tiny that we would not be able to observe them?" asked Misha. "Before the microscope invention no one knew that the world of microorganisms exists."

"The matter of the fact is that we are going to perceive the life of the Pulsar not by our senses but through spiritual communications," answered L.A. "Therefore, the physical boundaries having to do with the size of the object are not essential to us. But now it is your turn. You must either willingly agree to the experiment or decisively reject it. The success of the experiment mostly depends on your complete willingness to participate. The slightest hesitation may spoil everything. If you wish, your commitment is your belief. Please, take note once more that the experiment will in no way affect your health, if you agree to it with complete understanding and trust."

"I am ready and willing to join the experiment," said Sasha loudly.

"Me too." Misha was no less decisive.

"Then let us begin," concluded L.A.

Immediately assistants in white gowns appeared from nowhere. They started a lengthy procedure of attaching numerous metallic plates to different parts of the body and head. The armchairs were inclined in such a way that all the three participants were half lying down.

"All is ready," said the senior assistant finally. "Everybody, please calm down and relax and do not think of anything else. It will help if you close your eyes."

Misha closed his eyes and calmly stretched in the armchair. All was quiet .... But gradually the quietness changed to a sort of hum. Then the humming became louder and Misha thought that he had heard this low frequency sound before. The sound grew louder and louder. Then the sound started to be accompanied by some kind of images. First they looked like bursts of sharp red light. Then the central part of pulsating red light became whiter and the periphery grew darker. And the noise was almost gone. Finally, only a light circle remained, surrounded by darkness. "Light at the end of tunnel," flashed in Misha's mind. "But this is the very tunnel drawn in the famous picture by Ieronimus Bosch, which brings the souls of the dead to another world," he thought.

But he did not have time to comprehend in his semi-consciousness what was happening until he heard L.A.'s even and assured voice.

"We are proceeding to the Pulsar. The bright circle in front is the Pulsar, i.e. the neutron star," explained L.A.

The circle grew in size, acquiring a bluish shade. Finally, it covered everything and Misha saw a solid blue surface irradiating light.

"As you may see, the Pulsar's surface is solid."

"That is true, but it is absolutely smooth and shines as if it were recently polished," Sasha interrupted.

Indeed, the smooth and slightly dull surface looked like artificial ice evenly illuminated from beneath with soft blue light.

"Do not make haste with your conclusions. Let us carefully examine the surface and then you may propose your impressions and observations," said L.A. It seemed to Misha as if he were moving his head closer and closer to the smooth surface.

"We have just started to adapt our dimensions to the dimensions of possible inhomogeneities on the surface. It is as if we are getting smaller," sounded L.A.

The surface grew closer and larger. Misha started to distinguish the details. They seemed semi-transparent to him and were hardly visible. All of a sudden he could clearly see the whole tracery of the surface.

"I've just seen a beautiful tracery!" exclaimed Misha.

"The surface represents a complex bone lace," Sasha joined his friend.

"I would like you to be very attentive and observant and speak aloud of everything you see," said L.A.

"I see a tracery which consists of a great number of tiny lumps connected with each other by numerous threads. This complex picture reminds me of a neuron net which I not long ago saw in the book on physics. Can a star have the same structure? This is something similar to life!" started Sasha.

"We may find a new form of life in the star. The most important thing necessary to maintain life is the energy flow, which is fairly orderly. If we characterize this order as negentropy, or negative entropy, then the energy flow should possess a very high negentropy. The Pulsar has such energy. This is a thermal flow from inside the Pulsar. As the Pulsar's temperature is extremely high, the entropy of the flow is pretty low. The thermal energy flow is being irradiated into cosmic space. A part of this flow can be 'overcaptured' by complicated structures in order to sustain these structures for their further development. If the energy of the Sun is enough to maintain life on Earth, then the question arises: why is the Pulsar thermal flow not enough to form structures?"

"But the temperature on Earth is only 300 K while the Sun temperature is twenty times higher," said Sasha. "Which is why the energy of the Sun can be considered highly ordered."

"True," said L.A. "The contrast in entropies between the consumed and released energies in the life on Earth is drastic if we take into account that the energy consumption starts with photosynthesis in plants' chlorophyll.

"But life may also exist at a lower contrast of entropies. It is known, for example, that close to geothermal sources on the ocean's bottom certain bacterial species exist that sustain their life only due to the thermal flow from the source. This slight deviation from equilibrium proves to be sufficient to synthesize highly organized bacteria. I believe that the contrast between a very hot Pulsar's surface and the cold space may also be sufficient to sustain some kind of unusual life.

"I should draw your attention to the fact that any life is a permanent movement. Any living organism is a clot of a highly organized structure which sustains its existence only at the expense of perpetual metabolism or, to be more precise, perpetual exchange of matter and reconstruction of all destroyed structures. As soon as 'feed' stops, life terminates. Therefore, if we actually confronted a new form of life we should try to understand what kind of motion supports it."

"I've noticed some motion," exclaimed Misha.

"Look, over each lump which could be called a neuron there is some foggy cloud which produces drops. These drops precipitate on the hillock and then the liquid streams down. Something like micro-rain."

"This is something extraordinary," exclaimed L.A. "The human body has a special circulatory system which pumps blood and sustains metabolism. On our planet the atmosphere plays the same role. It develops clouds, transporting them over large distances, irrigates the earth while rivers, like man's blood vessels, remove the polluted water far to the sea. But here each micro-organism creates a small rain for itself. Such a version of 'blood circulation' can take place only under extraordinary physical circumstances. There should be such a substance that could be condensed at temperatures slightly lower than the temperature of the Pulsar's 'naked' surface. Each lump neuron slightly shields the heat flow from inside the Pulsar, and the temperature drop due to this shield is sufficient to condense vapours. All this reminds us of the 'Magic bird', a glass toy which constantly bends down 'to drink water' from the glass. From inside the leak-proof retort, the bird's body, the ether condenses in the cold beak of the bird when the latter cools it down due to water evaporation. The bird, with a heavy head, bends down and places its beak into the glass of water, the ether evaporates and is shifted into the body of the bird. Then the bird straightens up and swings with the pulse. The beak cools down and the bird 'drinks' again. But here on the Pulsar the same occurs locally under each neuron if this is really a neuron."

"But what about rain? Can we expect it here?" asked Sasha.

"It sure can rain," answered L.A. "I have just explained to you that the winds on the Earth are born due to local convection that entails the effect of negative viscosity when we have large-scale flows. They transport water vapours at large distances. Condensing, the vapours develop oversaturated clouds and rain starts. The same can happen here on the Pulsar. There is atmosphere here and the conditions for local convection are also satisfied because the Pulsar generates constant thermal flow from the internal layers to the surface. Also, there are conditions to stimulate rain from certain substances.

"Now let us consider this structure more in detail. If this very complicated web, which covers all of the surface, can indeed transmit some current signals between neurons, we certainly face a real neuron net. It is better to say that this is a huge superbrain which covers all the Pulsar surface. Let us try to understand how this superbrain can function and how it 'thinks'. Please, try to find some details."

"I have noticed very small strange sparks. I read that similar sparks on Earth are called 'Holly Elm fires'. But why do they appear here?"

"I believe," said L.A., "that they have the same origin as on Earth. Holly Elm fires are in fact the sparks of corona discharge. On Earth they appear when a very high electric field is produced by a thundercloud. By the way, the corona discharge you can hear and sometimes even see when you are staying under the high voltage electric net line. Here on the Pulsar a very strong electric field is produced due to the Pulsar rotation. A magnetic field is frozen in the Pulsar surface and its rotation automatically leads to the electric field build up. The sparks we see are in fact electron beams which are extracted from the surface and then accelerated by a very strong electric field."

"Look what a strange structure I have found," exclaimed Misha. "It looks like a great number of gratings arranged in such a way that brings to mind a giant chessboard. How could such a strange structure appear here?"

"This is indeed very surprising," exclaimed L.A. "Such a complicated structure could appear only when some kind of intellect is involved. Most likely that this intellect belongs to the Pulsar itself," said L.A. in a manner that definitely indicated that he started to believe that the Pulsar is an animated subject. "This structure looks as a very sophisticated one. This reminds me of the very radiotelescope that helped to discover the first Pulsar in 1967. Remember that it consisted of a number of properly phased antennae. What is the reason for such a structure here?

"We cannot disregard the fact that the nervous system of the Pulsar functions quite differently. Using controlled electron beams one can excite electromagnetic waves in the plasma surrounding the Pulsar and the antennae can receive these waves. This means that the bonds between neurons could be effected without 'wires', but be of electromagnetic nature. This type of bonds could be much more effective because the electromagnetic waves propagate with the speed of light.

"Besides, the phased antennae can receive very distant signals from space. For instance, they could be reflected signals initially emitted by the Pulsar. In this particular case, the Pulsar represents a huge radar. But there could also be very weak signals from far-off sources of radiowaves in space."

"Maybe the Pulsar watches other pulsars or even talks to them?" joked Sasha.

"Let me think a little. Your question, Sasha, could be not far from the truth. Obviously, if the Pulsar is a superbrain, then it has some objectives in its 'deliberations'. And it certainly has a need to communicate with the rest of the World. Such a Pulsar needs 'to speak' or 'to hear'. It can speak, i.e. create and emit information, by means of radiation which can be recorded on the Earth. The sparkles we see could be not just accidental but produced as a result of 'deliberate' control of electric conductivity of separate parts of the Pulsar's surface. This is the way the superbrain can 'express itself'. In the presence of a very complicated correlation between the emitted electron beams and radiation quanta which are excited by those beams, we cannot exclude the possibility of transmitting a huge amount of information by the Pulsar proper. It was due to this information that we could undertake our travel to the Pulsar. Therefore, we have proved that the Pulsar is most probably able 'to speak.' But how can it 'hear'?

"If these lattices are actually phased antennae, then the Pulsar can receive the same information that we can receive in our laboratory on Earth. This information is 'recorded' in the correlating bonds among radiation quanta. If this is true, the Pulsar is a superintellect which could directly communicate with other 'thinking objects' of the Universe."

"But does intellect exists beyond man?" asked Misha.

"Certainly," answered L.A. "To think that man is a supreme creature of Nature or God is to be extremely short-sighted. Nature is much richer than man and its intellect, as a whole, significantly exceeds the relatively modest intellectual abilities of a man."

"Then what is the relationship between man and the Pulsar? Maybe the Pulsar is a very Big Eagle that feeds on human thoughts."

"Strange as it may seem, we cannot exclude this possibility," answered L.A. "Indeed, if the Pulsar is a colossal superbrain, with a tremendous urge to perceive then it should try to master as much information as possible. Thinking creatures, like man, are an additional possibility for it to expand its intellectual abilities. It is obvious that the urge to perceive is immanent and should cause the Pulsar its information hunger. Even we, being mortals, demand 'bread and excitement', i.e. to be fed materialistically as well as informationally. It is obvious that the Superintellect's information hunger is even greater and, like a bee, it should collect 'knowledge nectar' from everywhere possible. Subject to a more active position, it can even stimulate the other thinking creatures to fill up its knowledge base as fast as possible."

"Then it could be said that the Earth is just an incubator of knowledge for the Superintellect?" asked Sasha.

"Why not?" replied L.A. "Naturally, we, human beings, have developed an exaggerated feeling of one-upmanship and often consider ourselves the top of the Universe. That is not true: the Universe must have goals and objectives that we cannot comprehend. So let us be more modest and be happy with what we have."

"L.A.!" the assistant shouted out. "Unfortunately, we had to wake you and your friends up. You slept so soundly that we started to worry about your health."

"Damn!" exclaimed L.A.

"As usual the dream stops at the most interesting stage. But cheer up, guys, we will undertake many other interesting journeys."

Appendix

Problem No. 1

We want to find how the elastic 'reflection' of an anisotropic point mass from a solid surface occurs. For an isotropic mass the law of reflection is very simple: the absolute speed of the particle is retained and the reflection angle is equal to the incidence one. For an anisotropic mass this is not true. Let's take a system of coordinates (x, y) where the x-axis lies along the plane of reflection and the y-axis is directed along the normal to this plane (Fig. A.1). Let us denote by α the angle between the normal and the magnetic field.

The particle momentum $\boldsymbol{p}$ can be resolved into longitudinal and lateral components: $\boldsymbol{p} = \boldsymbol{p}_\parallel + \boldsymbol{p}_\perp = m_\parallel \boldsymbol{v}_\parallel + m_\perp \boldsymbol{v}_\perp$.

Let's note once again that the longitudinal mass can simply be written as m. In addition we introduce the momentum components p_x, p_y:

$$p_x = p_\parallel \sin \alpha - p_\perp \cos \alpha$$
$$p_y = p_\parallel \cos \alpha + p_\perp \sin \alpha$$

Let us also introduce the particle energy for consideration:

$$\mathcal{E} = \frac{p_\parallel^2}{2m_\parallel} + \frac{p_\perp^2}{2m_\perp}$$

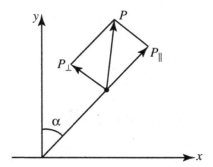

Fig. A.1   Momentum $\boldsymbol{p}$ decomposition into longitudinal $p_{\parallel}$ and lateral $p_{\perp}$ components.

In the usual isotropic mass case, an elastic collision corresponds to the conservation of the tangential momentum component p_x and a change in the sign of the normal component p_y.

In the case of anisotropic mass this recipe is not suitable: if we simply substitute $-p_y$ for p_y whilst maintaining the value of $p_x \neq 0$, the particle's energy will not be conserved. This can be checked by a simple calculation which we will skip. Therefore, such a conversion does not correspond to an elastic reflection.

What do we need for the reflection to become elastic? The condition of $p_x = $ const is obligatory as there are no reflection forces along the plane. As for the y-component, the following is true. The elastic collision of a particle should produce a fully reversible motion: first, the particle deforms the plane displacing it slightly below $y = 0$; then it performs the reverse motion, gaining a y-component of velocity in the opposite direction. Therefore, the condition for elastic reflection is based on the fact that p_x is maintained whereas the y-component v_y of the velocity changes sign. Using the above relations for energy and momentum, it is possible to express the particle energy in terms of p_x and v_y:

$$\mathcal{E} = \frac{1}{2m_*} \left(p_x^2 + m_{\parallel} m_{\perp} v_y^2 \right)$$

where $m_* = m_{\perp} \cos^2 \alpha + m_{\parallel} \sin^2 \alpha$. As we see, the particle energy is an even function of v_y, so if v_y is substituted for $-v_y$ at $p_x = $ const the energy is conserved. Thus, we have found the conditions for the elastic reflection of the particle from the plane: $p_x = \text{const}$, $v_y \rightarrow -v_y$.

Now let us consider some specific cases. The simplest case is $p_x = 0$ where the particle does not possess a tangential momentum component. The momentum in this case is directed along the y-axis, and it is along this direction that the particle with isotropic mass would have been moving. But if the anisotropy is high enough, $m_\perp \gg m_\parallel$, then at comparable values of $p_\parallel$ and $p_\perp$ the particle will move along the magnetic field because the longitudinal speed component $v_\parallel = p_\parallel/m_\parallel$ turns out to be much higher than the lateral one $v_\perp = p_\perp/m_\perp$. This is the motion of the freely falling ball that we found in the experiment 'on the Pulsar': $m_\perp \gg m_\parallel$. It can be shown that at $p_x = 0$ the relationship $p_y = m_\parallel m_\perp v_y/m_*$ is true. Therefore, in our case the elastic reflection with reverse polarity of v_y automatically means a change of polarity of p_y. The law of energy conservation in this case is not violated. Thus, if, for example, you hit a ball with a racket in the y direction, it will recoil with double the speed of the racket along the y axis, but when $m_\perp \gg m_\parallel$ the ball will fly almost along the magnetic field direction. It is in just this particular case that the resultant momentum would be truly directed along the y axis.

Now let us consider the example when $p_x \neq 0$. It would again be suitable to choose a case of reflection that drastically differs from a regular one with an isotropic mass. Let's consider $v_x = 0$ prior to the stroke, so that the particle drops vertically down to the horizontal plane $x = 0$. Before the particle strikes the plate we have

$$v_\perp^0/v_\parallel^0 = \tan\alpha, \quad v_y^0 = v_\parallel^0/\cos\alpha$$
$$p_x^0 = -v_\perp^0(m_\perp - m_\parallel)\cos\alpha$$

After the stroke the value of p_x is kept and v_y changes its polarity. Now we could try to find the expression for $v_\perp/v_\parallel$ after the stroke. To do so, we should express $v_\perp$ and $v_\parallel$ in terms of p_x, v_y and then substitute into the expression values of p_x, v_y after the stroke. After simple calculations, we can obtain an expression which at $m_\perp \gg m_\parallel$ reduces to a simple relation: $v_\perp/v_\parallel \simeq -\cot\alpha$. After the stroke by the wall the particle almost slides along the plane $y = 0$: the vertical speed component of the particle turns out to be significantly lower than the horizontal speed. Therefore, the elastic reflection of the particle from the plane at $p_x \neq 0$ and $m_\perp \gg m_\parallel$ is completely different from the isotropic mass particle reflection.

Problem No. 2

Reduce the problem of an elastic collision of an anisotropic mass point with a plane to the problem of an elastic collision of an isotropic mass point using the appropriate conversion of coordinates.

The idea is as follows. If $m_\perp \gg m_\parallel$ the particle is much more inclined to move in the longitudinal direction than in the lateral one. But if we introduce a new longitudinal length scale, then we could try to remove this difference. Let z be a coordinate along $\boldsymbol{B}$. Then we introduce a new coordinate $z' = \boldsymbol{a}z$ and, consequently, a new longitudinal speed $v'_\parallel = \boldsymbol{a}v_\parallel$.

If $\boldsymbol{a}$ is low, then minute displacements along z' with a proportionally reduced longitudinal component of speed will correspond to large displacements along the magnetic field. Now we can try to select the value of $\boldsymbol{a}$ in such a way that in the new longitudinal and old transverse coordinates, the particle behaves as if it has isotropic mass.

Let us return to Fig. A.1 and start with the simplest case of $p_x = 0$: when being reflected the particle repeats its trajectory in the reverse order. Note, that $p_\parallel$ and $p_\perp$ are related to each other by $p_\parallel \sin\alpha = p_\perp \cos\alpha$ (see Fig. A.1). But the value of $p_\parallel$ could be substituted for $p_\parallel = m_\parallel v_\parallel = \boldsymbol{a}^{-1}m_\parallel v'_\parallel$. We want the particle to be isotropic in the new variables i.e., we want the relationship between the longitudinal and lateral momentum components to be the same. For this to happen, the new longitudinal (dashed) momentum should look like $p'_\parallel = m_\parallel v'_\parallel$. Therefore, the particle under consideration (in the old system of coordinates) should satisfy the following relationship:

$$a^{-1}m_{\parallel}v'_{\parallel}\sin\alpha = p_{\perp}\cos\alpha = m_{\perp}v_{\perp}\cos\alpha$$

Let us note once more that in our transformation $p'_{\perp} = p_{\perp}$.

Now let us consider what happens with the system of coordinates (x, y) as the plane is compressed along $\boldsymbol{B}$ by the transition from z to z'. It is obvious that as $\boldsymbol{a}$ decreases, the axes x', y' start to turn, tending to become orthogonal to $\boldsymbol{B}$ in the case of $\boldsymbol{a} \to 0$. As we see, the axis x' turns clockwise, and the vector $\boldsymbol{v}'$ turns anticlockwise since the component $v'_{\parallel}$ decreases with $\boldsymbol{a}$. At a certain moment, $\boldsymbol{v}'$ is perpendicular to x'. The elastic reflection of an isotropic mass point from the plane at normal incidence corresponds to this very value of $\boldsymbol{a}$. Now let us find this $\boldsymbol{a}$. We denote by γ the angle between x' and the normal to $\boldsymbol{B}$ (see Fig. A.2). Since this angle is a result of the deformation of triangle OAB into triangle OAB' by reducing side AB' by a factor of $\boldsymbol{a}$ compared with side AB, $\tan\gamma = \boldsymbol{a}\tan\alpha$ (see Fig. A.2).

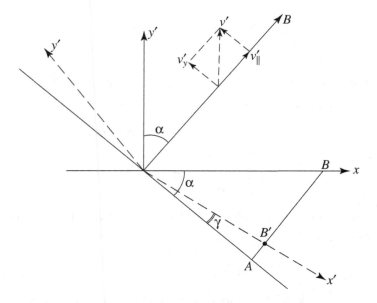

Fig. A.2   Transformation of the x, y axes into new axes x', y' when the picture plane is compressed along the $\boldsymbol{B}$-field direction.

The condition of for $\boldsymbol{v}'$ and x' to be perpendicular, as is seen from Fig. A.2, is: $v'_\perp \cos\gamma = v'_\parallel \sin\gamma$; or after expressing $\tan\gamma$ in terms of $\tan\alpha$ we obtain $\boldsymbol{a} v'_\parallel \sin\alpha = v_\perp \cos\alpha$. Comparing this relationship with the above condition of $p_x = 0$ in the old system of coordinates, we find

$$\boldsymbol{a}^2 = \frac{m_\parallel}{m_\perp}$$

Thus, if we compress the complete plane along the direction of $\boldsymbol{B}$ by a factor of $\boldsymbol{a} = \sqrt{m_\parallel / m_\perp}$ then a collision of an anisotropic particle with $p_x = 0$ with the plane x simply looks like the normal incidence of an isotropic mass particle on the plane x'.

It is easy to see that this transformation is also true in the general case of $p_x \neq 0$. Indeed, in the transformed system of coordinates the collision is as for an isotropic mass particle. The reflecting condition can be formulated as a combination of the conservation of the tangential component of momentum and a sign change in the normal component of the speed. Using Fig. A.2, we have:

$$m_\perp v'_\perp \cos\gamma - m_\perp v'_\parallel \sin\gamma = \text{const}$$
$$v'_\perp \sin\gamma - v'_\parallel \cos\gamma \rightarrow -(v'_\perp \sin\gamma - v'_\parallel \cos\gamma)$$

Now if we assume that $v'_\perp = v_\perp$, $\tan\gamma = \boldsymbol{a}\tan\alpha$, $\boldsymbol{a} = \sqrt{m_\parallel / m_\perp}$, $v'_\parallel = \boldsymbol{a} v_\parallel$, then the first condition can be written as

$$p_x = m_\perp v_\perp \cos\alpha - m_\parallel v_\parallel \sin\alpha = \text{const},$$

and the second as

$$v_y v_\perp \sin\alpha - v_\parallel \cos\alpha \rightarrow -v_y$$

Thus, we obtain the condition of elastic reflection of an anisotropic mass particle.

Problem No. 3

Reduce the task of the movement of an anisotropic mass particle under gravity to that for an isotropic mass particle using a transformation of coordinates.

This task is a bit more complicated than the previous one. Let us assume that a particle is under the gravity force $\boldsymbol{F} = m_\parallel \boldsymbol{g}$, where $\boldsymbol{g}$ is the free acceleration. Now the axes (x, y) should be chosen in such a way that x is oriented horizontally and y vertically. Then $\boldsymbol{g}$ will have only a y-component: $g_y = -g$. The particle motion can be decomposed into lateral and longitudinal components:

$$m_\perp \dot{v}_\perp = m_\parallel g_\perp, \quad m_\parallel \dot{v}_\parallel = m_\parallel g_\parallel$$

The dot here means time derivative, and the indices $\perp, \parallel$ are projections on the transverse and longitudinal directions, respectively.

Let us again perform a transformation of the plane by a compression along $\boldsymbol{B}$. In this case, $v_\parallel = v'_\parallel / \boldsymbol{a}$, so that the equations of motion, provided that $m_\parallel / m_\perp = \boldsymbol{a}^2$, can be written as follows:

$$m_\perp \dot{v}_\perp = m_\perp g_\perp, \quad m_\perp \dot{v}'_\parallel = \boldsymbol{a}^{-1} m_\parallel g_\parallel$$

It is clear that if we introduce a new acceleration due to gravity with the components $g'_\perp = g_\perp, g'_\parallel = \boldsymbol{a}^{-1} g_\parallel$, then in this new gravitational field the isotropic 'overweight' particle with mass $m_\perp$ will precisely simulate the movement of the initial anisotropic particle.

It is easy to check that in the new system the 'horizontal' coincides with the x' axis. In other words, the x' component of $\boldsymbol{g}'$ is equal to zero. We can check this statement by referring again to Fig. A.2, and assuming that g' has negative lateral and longitudinal components:

$$g'_\perp \cos \gamma - g'_\parallel \sin \gamma = 0$$

Substituting $g'_\perp = g_\perp$, $g'_\parallel = \boldsymbol{a}^{-1} g_\parallel$, $\tan \gamma = \boldsymbol{a} \tan \alpha$, we obtain:

$$g_\perp \cos \alpha - g_\parallel \sin \alpha = 0$$

The last statement is obvious since we have chosen the x-axis to be directed horizontally.

Therefore, if we turn the dashed plane (Fig. A.2) in such a way that the x'-axis becomes horizontal, then we can simulate the behavior of strongly non-isotropic particles in a gravitational field using isotropic mass particles. Now it is clear why the ball (Fig. 3.5) falling against the y-axis slides down, bouncing, and then continues to bounce on the horizontal plane.

Problem No. 4

Reduce the equations of ideal hydrodynamics for an anisotropic mass liquid in a gravitational field to isotropic form using the necessary change of the coordinate system.

Let the liquid be anisotropic, so that $\rho_\perp \gg \rho_\parallel$ where the indices $\perp, \parallel$ signify directions across and along the magnetic field, respectively. Let us introduce the coordinate z along the magnetic field direction. Then the Euler equations can be written as:

$$\rho_\parallel \frac{d}{dt} v_z = -\frac{\partial p}{\partial z} - \rho_\parallel g_z$$

$$\rho_\perp \frac{d}{dt} \boldsymbol{v}_\perp = -\boldsymbol{\nabla} p - \rho_\perp \boldsymbol{g}_\perp$$

and the condition for incompressible flow can be written as:

$$\frac{\partial v_\parallel}{\partial z} + \boldsymbol{\nabla}_\perp \boldsymbol{v}_\perp = 0$$

Here the derivative

$$\frac{d}{dt} = \frac{\partial}{\partial t} + v_z \frac{\partial}{\partial z} + \boldsymbol{v}_\perp \boldsymbol{\nabla}_\perp$$

We assume that the gravity force is determined by the value of $\rho = \rho_\parallel$.

The idea of the reduction of equations to the isotropic liquid equations is that by introducing a new scale along the z direction we can eliminate the difference in liquid behavior along and across the magnetic field. Let $v_z' = \boldsymbol{a} v_z$, $z' = \boldsymbol{a} z$; the dash labels new coordinates. If $\boldsymbol{a} \ll 1$, insignificant changes of the dashed variables will correspond to significant changes of v_z and z. In other words, the dashed variables 'slow down' the longitudinal process. If $\boldsymbol{a}$ is chosen correctly, we can expect the longitudinal and lateral motions to be of the same nature, i.e. the liquid becomes isotropic in these variables. Thus, let us change z and v_z for z' and v_z'. It is easy to see that total time derivative d/dt does not change its form; the equation of non-compression remains the same too, but the components of Euler's equation change:

$$\rho_\| \frac{1}{a^2}\frac{d}{dt}v_z' = -\frac{\partial p}{\partial z'} - \frac{1}{a}\rho_\| g_z$$

$$\rho_\perp \frac{d}{dt}v_\perp = -\nabla_\perp p - \rho + \rho_\| g_\perp$$

If we choose $a^2 = \rho_\|/\rho_\perp$ and introduce a new gravitational acceleration with longitudinal component $g_z' = g_z\sqrt{\rho_\perp/\rho_\|}$, the obtained equation becomes isotropic. In the new system of coordinates the new equations describe the flow of isotropic liquid with mass density $\rho_\perp$. This flow corresponds precisely with the flow of anisotropic liquid in the initial system of coordinates. If the relation $\rho_\perp/\rho_\| = b^2$ is valid, then $z' = z/b$, $v_z' = v_z/b$, i.e. we obtain the transformation that was considered in the main text. It was seen in the text how to use it, i.e. the liquid with its boundaries (vessel, bottom of the water reservoir, free surface) should be first compressed along the magnetic field direction B, decreasing the scale b times. Then one has to increase the longitudinal component of the gravity force b times. After that it is convenient to turn the obtained picture round in such a way that the force of gravity is directed downwards. The behavior of the 'overweight' liquid with density $\rho_\perp = b^{-2}\rho$ in this new geometry simulates the anisotropic liquid behavior.

Problem No. 5

Taking into account the electron's wave properties, estimate the dimensions of a hydrogen atom, the electron cloud 'stick' diameter in the magnetic field, and the magnetic field intensity B_0 above which the magnetic field can be considered superstrong.

The particle's wave properties are revealed by the fact that the particle's momentum p is related to the de Broglie wavelength $\lambda = h/p$. Here h is Planck's constant. If you try to localize a particle within distance Δx, the possible wavelength will be smaller than Δx; which is why the particle momentum uncertainty Δp appears.

As a result, we arrive at the famous uncertainty principle:

$$\Delta p \Delta x > \hbar$$

where $\hbar = h/2\pi$.

Using this ratio we can estimate all the required values.

The hydrogen atom's electron is in the proton's electric field with potential energy $U = -e^2/r$, where r is the distance to the proton. If the atom's characteristic radius is a_0, the potential energy is approximately equal to $U \simeq -e^2/a_0$. The kinetic energy $K = p^2/2m$ can be estimated with the help of the uncertainty principle:

$$K \simeq \frac{\hbar^2}{2ma_0^2}$$

It is known that in a Coulomb field the kinetic energy is half the potential energy, but with the opposite sign, so that $e^2/a \simeq \hbar^2/ma_0^2$. Thus we can find out the radius of a hydrogen atom:

$$a_0 = \frac{\hbar^2}{me^2} \simeq 0.53 \cdot 10^{-8} \text{ cm}$$

Now let us determine the electron cloud's size across a magnetic field B. To do so, remember that the Larmor radius $\rho_\perp$ is equal to:

$$\rho_\perp = \frac{mv_\perp}{eB} = \frac{p}{eB}$$

The momentum can be estimated with the help of the uncertainty relation:

$$p \approx \hbar/\rho_\perp$$

From these two relationships we obtain:

$$\rho^2 \simeq \frac{\hbar}{eB}$$

The stronger the magnetic field, the smaller the electron cloud radius $\rho_\perp$ is. When the latter becomes smaller than a hydrogen atom, the magnetic field can be considered to be superstrong. Thus, from the relationship

$$\rho_\perp^2 = \frac{\hbar}{eB} = a_0^2 = \frac{\hbar^4}{m^2e^4}$$

we obtain the value B_0 above which the magnetic field can be considered superstrong, i.e.

$$B_0 = m^2 e^3 \hbar^{-3} = 2.35 \cdot 10^9 \text{ Gauss}$$

On Earth, such fields are impossible. However, they can be even stronger on a pulsar.

Problem No. 6

Evaluate the transverse mass of an atom in a superstrong magnetic field. Let the longitudinal, i.e. regular mass, be equal to M, and the atomic number be equal to Z. This means that the nucleus has charge Ze, and Z electrons are 'rotating' around it. Let us assume for simplicity that the atom is of the same diameter, a as a hydrogen atom. If the atom moves across the magnetic field with velocity $v_\perp$, then the Lorentz force $ev_\perp B$ affects every electron in the atom. This force leads to the deformation of the electron cloud, and to an increase in the potential energy as the electrons are displaced from the attractive nucleus. Let us find the velocity $v_\perp^0$ at which the shell will be shifted a distance of the order of the initial size of the atom. This will happen when the Lorentz force becomes comparable with the force attracting the electrons to the nucleus, i.e.

$$ev_\perp^0 B \simeq Ze^2/a^2$$

At this velocity, the total potential energy of all the electrons will increase by $\mathcal{E}_z$, where $\mathcal{E}_z$ is equal to:

$$\mathcal{E}_z \simeq Z^2 e^2/a$$

At lower velocities, the corresponding energy will decrease with $v_\perp$ according to the equation:

$$\mathcal{E} = \mathcal{E}_z (v_\perp/v_\perp^0)^2$$

Substituting in the expressions for $\mathcal{E}_z$, a and $v_\perp^0$ we obtain the following value for the 'excess' energy:

$$\mathcal{E} \simeq mv_\perp^2 \left(\frac{B}{B_0}\right)^2$$

This expression should be compared with the kinetic energy:

$$K = \frac{M}{2}v_\perp^2$$

Summing the kinetic energy and the 'added' energy $\mathcal{E}$, which is also quadratic in velocity, we find the expression for the effective lateral mass:

$$M_\perp \simeq M\left\{1 + \frac{2m}{M}\left(\frac{B}{B_0}\right)^2\right\}$$

As we see, when $B \sim B_0$ the effect of mass anisotropy is very low. However, if $B/B_0 \geq 10^2$, the mass becomes highly anisotropic so that the lateral mass can considerably exceed the longitudinal one.

Problem No. 7

Estimate the content (in percent) of protons in the neutron star.

Unlike an atomic nucleus, a neutron star cannot have a high positive charge. It is, so to speak, quasi-neutral — if it does contain protons, the number of electrons should be approximately the same. Otherwise, an electric field of a very high energy would be formed. Thus, if n_p is the number density of protons and n_e the density of electrons, then $n_p \approx n_e$. A small net charge can appear if the number of electrons and protons isn't quite equal.

Let n be the number density of neutrons (the number of particles per cubic cm). We will now see that $n_p/n \ll 1$. The reason for this is again the famous uncertainty principle.

In a neutron star the average distance between neutrons is $\Delta x = n^{-1/3}$. Therefore, according to the uncertainty principle the neutrons each possess momentum $p \sim \hbar/\Delta x \sim \hbar n^{1/3}$. From this one can estimate the average neutron kinetic energy:

$$\mathcal{E} = p^2/2m_n \sim \hbar^2 m_n^{-1} n^{2/3},$$

where m_n is the mass of a neutron (we omitted the factor of $1/2$ to simplify the estimate).

All the above is also applicable to protons, and if their density is lower than the density of neutrons, their kinetic energy is low. As for the electrons, their average momentum p_e can be estimated with the help of the uncertainty principle:

$$p_e \sim \hbar n_e^{1/3}$$

where n_e is the electron density. In a neutron star this momentum is high enough that the electrons become strongly relativistic. This means that their energy should be estimated using the relationship:

$$\mathcal{E}_e = p_e c$$

where c is the speed of light.

The electron energy is high, so it is unfavorable for them to be present in the star. In other words, their density should be much lower than the neutron density; but due to quasi-neutrality the proton density n_p is equal to the electron density. Therefore, the proton density should be only a small fraction of the neutron density.

Neutrons can decay into an electron-proton pair (emitting an anti-neutrino), and a similar reaction can turn an electron and a proton back into a neutron. Equilibrium between these two processes is achieved when the electron energy becomes equal to the neutron kinetic energy (the rest energy of the electron, and the kinetic energy of the proton are too small to play any significant role in the calculation).

Thus, from the condition $\mathcal{E} \sim \mathcal{E}_e$ we obtain:

$$(n_p/n)^{1/3} = (n_e/n)^{1/3} \simeq \frac{\hbar}{m_n c} n^{1/3}$$

where $m_n = 1.6 \cdot 10^{-27}$ kg is the neutron mass, $c = 3 \cdot 10^8$ m sec^{-1} is the speed of light, and the Planck constant $\hbar \simeq 10^{-34}$. When the neutron density is of the order of that of an atomic nucleus, $n^{1/3} \simeq 10^{13}$, the above estimate gives $(n_e/n)^{1/3} \simeq 0.2$. Thus, the proton density in a neutron star is about two orders of magnitude lower than the neutron density.

Index